T4-BCI-462

PLANT FOSSILS IN GEOLOGICAL INVESTIGATION
The Palaeozoic

ELLIS HORWOOD SERIES IN GEOLOGY

Editors: D. T. DONOVAN, Professor of Geology, University College London, and J. W. MURRAY, Professor of Geology, University of Exeter

This series aims to build up a library of books on geology which will include student texts and also more advanced works of interest to professional geologists and to industry. The series will include translation of important books recently published in Europe, and also books specially commissioned.

RADIOACTIVITY IN GEOLOGY: Principles and Applications
E. M. DURRANCE, Department of Geology, University of Exeter
FAULT AND FOLD TECTONICS
W. JAROSZEWSKI, Faculty of Geology, University of Warsaw
A GUIDE TO CLASSIFICATION IN GEOLOGY
J. W. MURRAY, Department of Geology, University of Exeter
THE CENOZOIC ERA: Tertiary and Quaternary
C. POMEROL, University of Paris VI
Translated by D. W. HUMPHRIES, Department of Geology, University of Sheffield, and E. E. HUMPHRIES
Edited by D. CURRY and D. T. DONOVAN, Department of Geology, University College London

BRITISH MICROPALAEONTOLOGICAL SOCIETY SERIES

This series, published by Ellis Horwood Limited for the British Micropalaeontological Society, aims to gather together knowledge of a particular faunal group for specialist and non-specialist geologists alike. The original series of Stratigraphic Atlas or Index volumes ultimately will cover all groups and will describe and illustrate the common elements of the microfauna through time (whether index or long-ranging species) thus enabling the reader to identify characteristic species in addition to those of restricted stratigraphic range. The series has now been enlarged to include the reports of conferences, organized by the Society, and collected essays on specialist themes.
The synthesis of knowledge presented in the series will reveal its strengths and prove its usefulness to the practising micropalaeontologist, and to those teaching and learning the subject. By identifying some of the gaps in this knowledge, the series will, it is believed, promote and stimulate further active research and investigation.

PALAEOBIOLOGY OF CONODONTS
Editor: R. J. ALDRIDGE, Department of Geology, University of Nottingham
CONODONTS: Investigative Techniques and Applications
Editor: R. L. AUSTIN, Department of Geology, University of Southampton
FOSSIL AND RECENT OSTRACODS
Editors: R. H. BATE, British Museum Natural History, London, and Stratigraphic Services International, Guildford, E. ROBINSON, Department of Geology, University College London, and L. SHEPPARD, British Museum Natural History, London, and Stratigraphic Services International, Guildford
NANNOFOSSILS AND THEIR APPLICATIONS
Editor: J. A. CRUX, British Petroleum Research Centre, Sunbury-on-Thames, and S. E. VAN HECK, Shell Internationale Petroleum, Maatschappij
A STRATIGRAPHICAL INDEX OF THE PALAEOZOIC ACRITARCHS AND OTHER MARINE MICROFLORA
Editors: K. J. DORNING, Pallab Research, Sheffield, and S. G. MOLYNEUX, British Geological Survey, Nottingham
MICROPALAEONTOLOGY OF CARBONATE ENVIRONMENTS
Editor: M. B. HART, Department of Geological Studies, Plymouth Polytechnic
A STRATIGRAPHICAL INDEX OF CONODONTS
Editors: A. C. HIGGINS, Geological Survey of Canada, Calgary, and R. L. AUSTIN, Department of Geology, University of Southampton
STRATIGRAPHICAL ATLAS OF FOSSIL FORAMINIFERA, 2nd Edition
Editors: D. G. JENKINS, The Open University, and J. W. MURRAY, Department of Geology, University of Exeter
A STRATIGRAPHICAL INDEX OF CALCAREOUS NANNOFOSSILS
Editor: A. R. LORD, Department of Geology, University College London
MICROFOSSILS FROM RECENT AND FOSSIL SHELF SEAS
Editors: J. W. NEALE, and M. D. BRASIER, Department of Geology, University of Hull
THE STRATIGRAPHIC DISTRIBUTION OF DINOFLAGELLATE CYSTS
Editor: A. J. POWELL, British Petroleum Research Centre, Sunbury-on-Thames
OSTRACODA
Editors: R. C. WHATLEY and C. MAYBURY, Department of Geology, University College of Wales

ELLIS HORWOOD SERIES IN APPLIED GEOLOGY

Published by Ellis Horwood Limited for the Institution of Geologists, Burlington House, Piccadilly London W1V 9AG

The books listed below are motivated by the up-to-date applications of geology to a wide range of industrial and environmental factors; they are practical, for use by the professional and practising geologist or engineer in the field, for study, and for reference.

A GUIDE TO PUMPING TESTS
F. C. BRASSINGTON, North West Water Authority
PLANT FOSSILS IN GEOLOGICAL INVESTIGATION: The Palaeozoic
Editor: C. J. CLEAL, National Museum of Wales
QUATERNARY GEOLOGY: Processes and Products
JOHN A. CATT, Rothamsted Experimental Station, Harpenden
NON-CONVENTIONAL METHODS IN GEOELECTRICAL PROSPECTING
MARK GOLDMAN, Petroleum Infrastructure Coporation Ltd, Holon, Israel
TUNNELLING GEOLOGY AND GEOTECHNICS
Editors: M. C. KNIGHTS and T. W. MELLORS, Consulting Engineers, W. S. Atkins & Partners
PRACTICAL PEDOLOGY: Manual of Soil Formation, Description and Mapping
S. G. McRAE, Department of Environmental Studies and Countryside Planning, Wye College (University of London)
STANDARDS FOR AGGREGATES
D. C. PIKE, Consultant in Aggregates, Reading
LASER HOLOGRAPHY IN GEOPHYSICS
S. TAKEMOTO, Disaster Prevention Research Unit, Kyoto University

PLANT FOSSILS IN GEOLOGICAL INVESTIGATION

The Palaeozoic

Editor
CHRISTOPHER J. CLEAL B.Sc., Ph.D.
National Museum of Wales

ELLIS HORWOOD
NEW YORK LONDON TORONTO SYDNEY TOKYO SINGAPORE

3.22

First published in 1991 by
ELLIS HORWOOD LIMITED
Market Cross House, Cooper Street,
Chichester, West Sussex, PO19 1EB, England

A division of
Simon & Schuster International Group
A Paramount Communications Company

Typeset in Times by Ellis Horwood Limited
Printed and bound in Great Britain
by Hartnolls, Bodmin, Cornwall

British Library Cataloguing-in-Publication Data

Plant fossils in geological investigation: The Palaeozoic. —
(Ellis Horwood series in applied geology)
I. Cleal, Christopher J. II. Series.
561

ISBN 0–13–680877–8

Library of Congress Cataloging-in-Publication Data

Plant fossils geological investigation: the Palaeozoic / editor, Christopher J. Cleal.
p. cm. — (Ellis Horwood series in applied geology)
Includes bibliographical references and index.
ISBN 0–13–680877–8
1. Paleobotany — Paleozoic. 2. Paleontology, Stratigraphic. 3. Paleogeography.
4. Paleoclimatology. I. Cleal, Christopher J., 1951– . II. Series.
QE15.P58 1991
560′.172–dc20

91–29159
CIP

List of contributors

Dr R. Bateman
Department of Paleobiology, National Museum of Natural History, Smithsonian Institution, Washington DC, USA.

Mr C. Berry
Department of Geology, University of Wales College of Cardiff, Cardiff CF1 3YE, UK.

Dr C. J. Cleal
Department of Botany, National Museum of Wales, Cardiff CF1 3NP, UK.

Dr D. Edwards
Department of Geology, University of Wales College of Cardiff, Cardiff CF1 3YE, UK.

Dr B. A. Thomas
Department of Botany, National Museum of Wales, Cardiff CF1 3NP, UK.

Table of contents

Preface

The idea for this book developed gradually over the last decade, as a string of textbooks on plant fossils were published giving, at least in my view, a biased account of their study. I was originally trained as a geologist, and found it normal to consider plant fossils in a geological context. Looking at much of the published literature, however, the newcomer would be forgiven for thinking that they were only of interest to botanists. By attempting to counter-balance this view, this book itself is biased. I make no apologies for this, but ask it to be viewed in the context of the rest of the recent palaeobotanical literature.

Although I have been responsible for writing part of the book, I am grateful to Dianne Edwards, Chris Berry, Barry Thomas and Richard Bateman for adding their particular expertise to some of the chapters. I would also like to thank John Cleal, for his help in preparing many of the diagrams and charts. Professor Bob Wagner (now of the Cordoba Botanical Gardens in Spain), for many years and against the odds, kept alive in Britain the geological study of plant fossils. Without him, this book would never have been written. As my PhD supervisor, many years ago, he encouraged a thoroughly healthy approach to the subject, guiding me in the direction of biostratigraphy and palaeogeography, as well as the more traditional pursuits of cuticle and frond architecture analysis.

Professor T. L. Phillips (University of Illinois, Urbana), Dr M. G. Bassett (National Museum of Wales, Cardiff) and Dr A.C. Scott (Royal Holloway & Bedford New College, London) are acknowledged for permission to reproduce some of the illustrations.

Finally, I must thank my wife Zoë, for her support and patience whilst working on this project. Without her encouragement, sympathy and cups of tea, I doubt if it would ever have been completed. Perhaps most significant was her willingness to take our two young daughters out for walks, often in less than pleasant weather, so that they were not tempted to come and 'play on daddy's computer'.

C. J. Cleal
June 1991

1

Introduction

C. J. Cleal

Plant macrofossils are a source of information relevant to both the geological and biological sciences. However, a newcomer to their study who consulted most recent English-language textbooks on the subject might be misled into thinking that they were only useful for understanding plant palaeobiology and evolution. Their titles certainly give this impression: *Evolution and plants of the past* (Banks 1972), *Paleobotany. An introduction to fossil plant biology* (Taylor 1981), *Paleobotany and the evolution of plants* (Stewart 1983), *The evolution and palaeobiology of land plants* (Thomas & Spicer 1986). This impression is confirmed by their contents, which are almost exclusively concerned with the morphology and evolutionary position of the extinct plants which produced the fossils. No criticism of these books is intended, as far as they go — they all provide excellent accounts of plant palaeobiology — but they do give a rather distorted picture of the range of information that plant macrofossils can provide. The only notable exception is Meyen's (1987) book *Fundamentals of palaeobotany* (translated from the Russian), which, in addition to providing a wide-ranging account of plant palaeobiology, has chapters devoted to the role of plant fossils in the Earth sciences. The palaeogeography section of this book is particularly useful, but those covering palaeoecology and biostratigraphy are essentially discussions on the principles and problems of the use of plant macrofossils in these fields. For instance, there is little detailed information on the biozones that have been established for such fossils.

There have been books published in recent years covering the use of plant macrofossils in geological work. Probably the most widely quoted is Krassilov's (1975) account of plant palaeoecology. This is undoubtedly a primary reference in the field, but is not the easiest to use in practice, largely because of the rather convoluted nomenclature that he develops for different taphonomic processes and assemblage-types. Another important book, edited by Dilcher & Taylor (1980), discusses aspects of plant biostratigraphy. There are a number of outstanding papers in this book, such as Banks' account of Silurian and Devonian biostratigraphy. However, it is very much a collection of isolated papers covering certain specific topics, rather than an attempt at a comprehensive review of the subject. Finally, it is worth noting the book on plant palaeogeography by Vakhrameev *et al.* (1978 — a German translation of an original Russian text). This provides a vast array of data on the geographical distribution of plant macrofossils, and is particularly useful for its coverage of Russian and Chinese assemblages, although it does not cover America, Africa or Australia. Clearly, no single published account covers all aspects of the geological application of plant macrofossils, and it is to fill this gap that this book has been compiled.

1.1 SCOPE OF THIS BOOK

This book discusses how terrestrial plant macrofossils can assist in the palaeogeographical, biostratigraphical and palaeoecological analysis of Silurian to Permian strata throughout the world. The aim has been to demonstrate the sort of resolution that they can provide, and in what facies they can be used. Because of the space available, it has been necessary to restrict the amount of raw data given on the fossils. Documenting, even at a superficial level, all of the taxa mentioned would take up at least ten times as many pages as are available. At least in part, therefore, this should be regarded more as a source-book, guiding the reader towards the literature in which the appropriate descriptions and illustrations of the fossils can be found.

For several reasons, the discussion has been limited to terrestrial plant macrofossils. Marine plant macrofossils are mainly the remains of calcareous algae (e.g. dasyclads, corallinids, solenoporids), whose study by thin sectioning approaches limestone sedimentology and petrography more closely than palaeobotany in its traditional sense (for a discussion of some aspects of this field, see Flügel 1977). The geological value of plant microfossils (e.g. pollen, spores, algal cysts) is undeniable, and has recently been discussed by Traverse (1988). However, they have to be studied using complex laboratory techniques, in which most general geologists will be unwilling to become involved. Throughout the rest of this book, therefore, wherever *plant fossils* are mentioned, they may be assumed to be *terrestrial plant macrofossils*.

When originally conceived, it was intended that this book would cover the geological use of plant fossils throughout the stratigraphic column. Various potential authors for chapters dealing with the Mesozoic and Cenozoic were approached, but it soon became clear that such fossils were rarely used for geological work in these strata. As even a brief review of such work in the Palaeozoic was filling the space being made available by the publishers, it was decided to restrict the scope of the book. Such a review is obviously of most interest to those geologists working in the Palaeozoic, but it is hoped that it will also have a wider significance, providing a model for how plant fossils may be used for geological work in strata of any age.

1.2 TECHNIQUES

1.2.1 Field techniques

Since this book is mainly targeted at the geologist, it should hardly be necessary to elaborate on the methods of collecting fossils in the field. Plant fossils do not differ significantly from any other palaeontological material, and their collection should be done in the same way. If further information is required, the reader is guided towards

Tucker (1982) for a general account of field methods, and Scott & Collinson (1983) for one more specifically directed to plant fossil collection.

Perhaps just three points should be emphasized. Firstly, plant fossils tend to be more delicate than many animal fossils, especially shelly ones. Consequently, more care needs to be taken in packing and transporting them from the field. They are best wrapped in several layers of newspaper and, in some cases, an inner layer of tissue paper (soft toilet paper works very well). Also, great care should be taken when specimens are stacked on top of one another, as this can cause damage to the fossil surfaces if insufficient packing has been used.

Secondly, evidence of the association of fossils together in a bed can be important. As in palaeozoology, it can provide important palaeoecological data. However, it may also give some indication of what plant organs were derived from the same plant; a seed and leaf associated closely in the same bed may indicate that they originated from the same plant (although not always — see Ferguson 1985).

Finally, every effort should be made to keep both the part and counterpart of fossils, particularly adpressions. When the rock is split and reveals an adpression, most information appears to be present on the piece showing the phytoleim (i.e. the carbonized plant tissue - see 2.1.3) known as the part. However, important details may also be present on the other piece (or the counterpart). This may not always be evident in the field, and is only discovered on further investigation of the fossil back at the laboratory.

1.2.2 Laboratory techniques

For the examination of plant fossils in the laboratory, the best tools are a good pair of eyes and a good light-source! However, this is normally best supplemented by at least a hand-lens; the author has found a 3 cm diameter lens with a magnification of about ×10 the most useful. A binocular dissecting microscope, with a magnification up to about ×25, can also prove invaluable. When using the latter, either a single powerful light or fibre-optic lighting is needed.

If there is good contrast between the fossil and the matrix, this is all that is needed. However, it is sometimes necessary to enhance the contrast before all the details can be clearly seen. Traditionally, this was done by flooding the surface of the fossil with xylene or some other volatile organic liquid. This produces quite satisfactory results, but care has to be taken in some cases that the liquid does not damage the matrix and loosen the fossil; it is also far from pleasant to spend any time breathing in the fumes produced by these liquids (there may also be a minor health hazard). Particularly if the fossil shows any surface topography, an alternative approach is to cover the specimen with a thin coat of ammonium chloride. This is a popular method with palaeozoologists. However, it is a most difficult technique to master, especially with larger specimens, over which it is difficult to get an even coating.

Recently, a third method has been developed — cross-polar lighting (Schaarschmidt 1975). It requires a piece of polarizing film over the light-source, and a rotatable polarizing filter on the objective lens of the microscope. By rotating the latter filter, the contrast can be variably enhanced to optimize results for both direct observation and photography. The results are as good as flooding the surface with organic liquids, but have the advantage of not presenting any risk to the specimen, not producing unpleasant fumes, and providing greater control over the degree of contrast. Even quite large specimens can be photographed by this method, using a flash for the light, although to get the best results with flash requires considerable experience. The main drawback is the reduction in the strength of the light, as it passes through the filters. For most practical purposes, a strong halogen lamp is needed, which is expensive. Nevertheless, if the equipment can be made available, the author can confidently recommend the method from his own experience.

If the fossil lies flat on the plane of fracture of the rock and is more or less two-dimensional, it should require little further preparation. Quite often, however, it does not lie flat and needs to be developed. The technique known as *degagement* (Leclercq 1960) involves the careful physical removal of the matrix overlying the fossil. For fine work, the best tools are a small hammer and fine steel needles (old-fashioned gramophone needles fixed in a small chuck are ideal, if they can be obtained). For more vigorous work, an electric engraving tool or a dentist's drill produces good results. The main drawback of degaging is that it can leave rather ugly scars on the matrix, which do not enhance the photogenic appeal of the specimen. This is, however, more of an aesthetic problem and, if carefully done, degaging can reveal much scientifically important data.

Palaeobotanists have developed numerous other techniques to maximize the information that can be derived from plant fossils. For instance, carbonized adpressions (see section 2.1.3) can be subjected to a variety of maceration techniques, to extract cuticles or to uncompress the fossil partially (e.g. Halle 1933, Cleal & Zodrow 1989). Some petrifactions can also be macerated, or can be sectioned to reveal anatomical detail using acetate peels or petrological sectioning methods (Joy *et al.* 1956). In recent years, advances in microscopical techniques have revolutionized the way palaeobotanists examine their fossils, particularly with scanning and transmission electron microscopes. Most textbooks describe these techniques, Taylor (1981) and Stewart (1983) being particularly helpful.

This type of work tends, however, to be time consuming, both in learning the techniques and applying them in practice, and they are therefore impracticable for the type of geological study with which this book is concerned. Such study depends on the identification of as many specimens as possible, in order to determine the distribution of the taxa, in time, space or ecological setting. In some cases, such work cannot be avoided; many petrifactions can only be identified reliably by sectioning, which is why they are rarely used in geological work. Where the fossils can be identified without recourse to such techniques, however, they are best avoided, at least in the first phase of the study. To attempt, say, cuticle preparations on several hundred specimens in a biostratigraphical study would be prohibitive, especially if the work was being done in a commercial setting. This is not to say that the geologist should not find out as much as he can about the fossils, but it has to be borne in mind that the results are likely to be of more botanical than geological significance.

1.2.3 Conservation

This term is used to cover two quite distinct concepts — site conservation and specimen conservation. The former lies outside of the scope of this book and will not be dealt with further (for a discussion on this subject, see Cleal 1988). The problem of specimen conservation, on the other hand, is a matter which any geologist handling plant fossils will eventually have to address.

It is strongly advised that any plant fossils uncovered in a geological investigation should eventually find their way to a museum collection. This should be a museum with at least a geological department, and preferably one with staff experienced in handling plant fossils. If this procedure is adopted, the long-term conservation of the specimen can be left in the hands of the museum specialists. The techniques behind the long-term conservation of plant fossils often require a great deal of experience (Collinson 1987), and is best not attempted by the non-specialist. All that the geologist should really be concerned with is the time of his stewardship of the material, between its extraction from the ground and its presentation to the museum.

The geologist may best ensure the longevity of the specimen by a number of *don'ts* rather than *dos*. In particular, try to avoid coating the specimen with varnish, as was popular some years ago. It can enhance the aesthetic appeal of the fossil, but will ruin any surface features or cuticles (Shute & Cleal 1987). For certain Cenozoic fossils preserved in very soft rock matrix, coating can be an unavoidable evil if they are to be retained intact, but this is rarely if ever a problem with Palaeozoic fossils. Other don'ts include not packing the specimens too closely together (preferably, they should be one to a specimen tray), not storing them in very damp conditions, and not losing the locality labels.

1.3 NAMING PLANT FOSSILS

Plant fossils are usually named using Linnaean-style binomial nomenclature, similar to that used for living plants. The International Code of Botanical Nomenclature (ICBN) provides the stabilizing framework to this system, the most important factors being the fixing of species to a type specimen (holotype), and the maintenance of chronological priority for the names. The problem is that fossils are not plants; they are dead pieces of carbon or other mineral, or even just empty spaces (decarbonized fossils — see section 2.1.3). Even assuming the fossils can be broadly equated with a living plant, the system of nomenclature enshrined in the ICBN does not fully take into account the time factor, and thus the dynamics of evolution. Alternative systems of nomenclature have been proposed which get over many of these problems (e.g. Hughes 1989), but the Linnaean system still holds sway with the vast majority of palaeobotanists.

One of the few problems specific to palaeobotany which the ICBN attempts to address is the fragmentary nature of fossils. Most specimens represent only part of the plant from which they were derived, perhaps a seed, a part of a stem or of a leaf. A goal for many palaeobotanists is to try to reconstruct most or all of a plant from these fragments, and there have been some remarkably successful attempts (e.g. Barthel 1968, Kerp 1988, Poort & Kerp 1990). However, these are by far the exceptions rather than the rule; the vast majority of plant fossils are the remains of an isolated plant organ (or even part of an organ) whose relationship to the rest of the parent plant is unknown.

To try to overcome this problem, the ICBN invokes the concept of *form-genus*, which allows different parts of the same plant to be assigned different taxonomic names. One of the most widely quoted examples used to illustrate this is the arborescent lycophyte, whose trunk is referred to the form-genus *Lepidodendron*. A typical reconstruction is shown in Fig. 1.1, together with four parts of the plant which are frequently found as fossils. The reproductive cones (or strobili) are referred to the form-genus *Flemingites* (1), the leafy shoots and trunk to *Lepidodendron* (2, 3) and the rooting structures to *Stigmaria* (4).

It is even possible to assign the same plant organ from the same parent plant, but preserved in different ways (e.g. adpression and petrifaction), to different form-genera. This is allowed because of the problems of correlating fossils preserved in different ways (Mosbrugger 1983, Galtier 1986). In recent years, such taxonomic distinction has become unpopular, as part of the 'new' holistic approach to palaeobotany. In principle, this is laudable; most people would agree that the fewer taxa there are, the better. In practice, however, great care has to be taken when trying to use the same taxon for adpressions and petrifactions. There is the obvious problem of the different data sets presented by each preservational state (see section 2.1.3 above). Also, different preservational states often developed in different habitats, and may thus represent quite different assemblages of plants.

A newcomer to the subject might find this system of form-genera unwieldy, and indeed it tends to give a false impression as to the diversity of an assemblage (e.g. see section 2.6); a species list of a dozen names may in fact only represent three or four original plant species. It is also not always clear from the name itself what plant organ is being referred to (to help the reader overcome this last problem, the type of plant organ that each name refers to will be specified in the taxonomic index provided at the end of the volume). There are other problems with the form-genus concept currently defined by the ICBN (Cleal 1986, Visscher *et al.* 1986), but these mainly concern how a form-genus is assigned to higher-ranked taxa (families, orders, etc.) and are thus of less immediate interest to the geologist wishing to name plant fossils. For most geological purposes, the form-genus provides a reasonably satisfactory tool for recording the distribution of the plant fossils.

1.4 WHAT PLANT FOSSILS ARE GEOLOGICALLY USEFUL?

In principle, any plant fossil can give geologically useful information but, in practice, some are more useful than others. There are two main criteria in determining their usefulness.

1. It is best if the fossils represent organs that were produced abundantly by commonly occurring plants. Most geological work where plant fossils have a role depends on their patterns of distribution, which can only be determined reliably if the fossils occur abundantly. A biostratigraphic model based on uncommon species is rarely robust in practice.
2. The fossils should show sufficient morphologic characteristics to allow easy and accurate identification. This should preferably not require involved laboratory

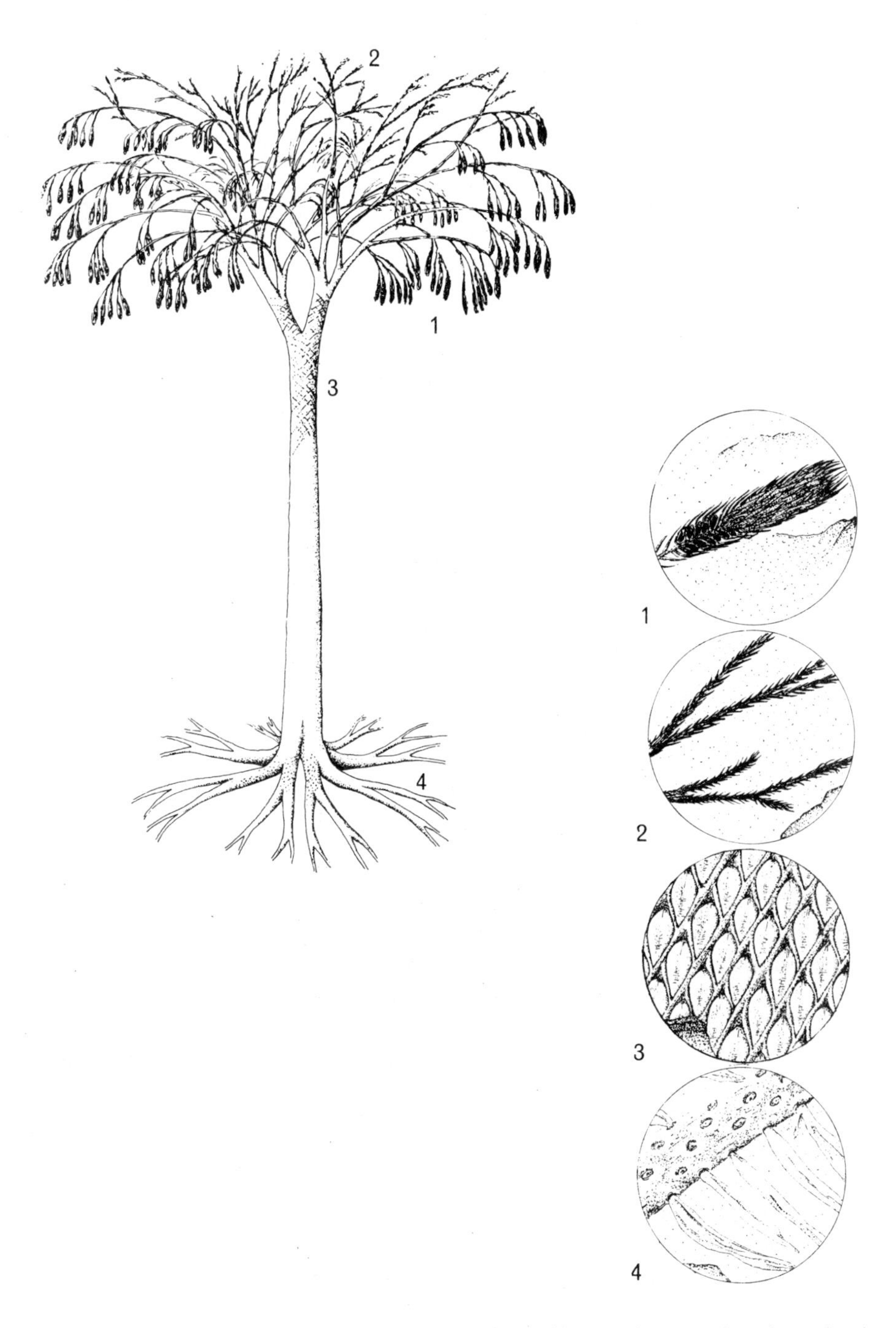

Fig. 1.1 — An example of a reconstructed Late Carboniferous arborescent lycophyte, showing four of its main component parts as found in the fossil record. 1, heterosporous strobilus (form-genus *Flemingites*); 2, foliage (form-genus *Cyperites*); 3, trunk with leaf scars (form-genus *Lepidodendron*); 4, rooting structures (form-genus *Stigmaria*). Reproduced from Cleal (1976, p. 19).

preparation (e.g. cuticle work, thin sectioning), which is often impracticable in routine geological investigation (see also section 1.2.2).

During the Silurian and Devonian, many of the plants were relatively small, herbaceous forms and so the fossils often represent more or less complete organisms (e.g. Fig. 1.2). In higher strata, however, arborescent forms dominate the fossil

Fig. 1.2 — *Steganotheca striata* (holotype). The remains of a small, herbaceous plant, typical of those represented in the Upper Silurian and lowermost Devonian. National Museum of Wales Specimen No. 69.64G.32b. From Capel Horeb, near Ludlow, Shropshire, UK (Ludfordian).

record. Here, foliage of ferns, progymnosperms and pteridosperms are geologically the most useful fossils (e.g. Figs 1.3, 1.4); they were produced abundantly by the plants, and have many gross-morphological characters to help in identification (e.g. pinnule shape and attachment, vein pattern, frond architecture). Sphenophyte foliage can also be useful (Fig. 1.5), although it usually shows fewer gross-morphological characters. Fossils of simple-leaved gymnosperms (e.g. cordaites, glossopterids, gigantopterids) also tend to have fewer morphological characters (Fig. 1.6). They have been used to good effect where they dominate the fossil assemblages, but the problem is that their identification has really to be backed up by an investigation of their cuticles, which may not be practicable for routine geological work.

Plant stems and roots frequently occur as fossils, but rarely show sufficient morphological characters on which to base an identification; it is often not possible to determine whether they originated from a gymnosperm, a pteridophyte or, in some

Fig. 1.3 — *Senftenbergia plumosa*. Part of a fern frond. National Museum of Wales, Specimen No. 21.43G.73. From Pendleton Pit, near Manchester, UK (?Duckmantian).

Fig. 1.4 — *Laveineopteris tenuifolia*. Part of a pteridosperm frond. Saarbrüacken Geological Museum, Specimen No. C/2087. From Seam 13, Saarland, Germany (Bolsovian).

Fig. 1.5 — *Sphenophyllum majus*. An example of sphenophyte foliage. National Museum of Wales, Specimen No. 22.11G.108 (David Davies Coll.). From No.3 Rhondda Seam, Gilfach Goch, South Wales, UK (Bolsovian).

Fig. 1.6 — *Palaeovittaria* sp. An example of foliage from a simple-leafed gymnosperm. National Museum of Wales, Specimen No. 77.25G.4. From the Lower Gondwana of India (Lower Permian).

cases, even an alga. Notable exceptions are the stems of arborescent lycophytes, on which leaf-bases are still preserved (e.g. Fig. 1.7). The shape of the leaf-base, the

Fig. 1.7 — *Sigillaria* sp. Part of a stem of an arborescent lycophyte. National Museum of Wales, Specimen No. G.1649. From Coalbrookdale, Shropshire, UK (?Bolsovian).

shape and position of the leaf-scar, and the position of parichnos and ligule marks are characters of taxonomic value in these fossils. Some sphenophyte stems and pith casts (Fig. 1.8) also have some taxonomically useful characteristics (e.g. node morphology, branch scars), but they tend to be more difficult to interpret than the characters available in the lycophytes.

For those concerned with the palaeobiology of the parent plants, the fructifications are normally regarded as the most significant. However, they are usually rarer as fossils than the foliage and stems, and are thus less useful for geological work. Also, unless they are anatomically preserved, they tend to be difficult to interpret; the gross morphology of a seed or cone is rarely enough to allow an accurate taxonomic determination.

1.5 THE GEOLOGICAL ROLE OF PLANT FOSSILS

The value of plant fossils for geological work depends mainly on their pattern of distribution, which in turn depends, at least in part, on the original distribution of the parent plants. The distribution of the plants would vary in space, time and habitat, and so the main uses of the fossils tend to be in palaeogeographical, biostratigraphical and palaeoecological work.

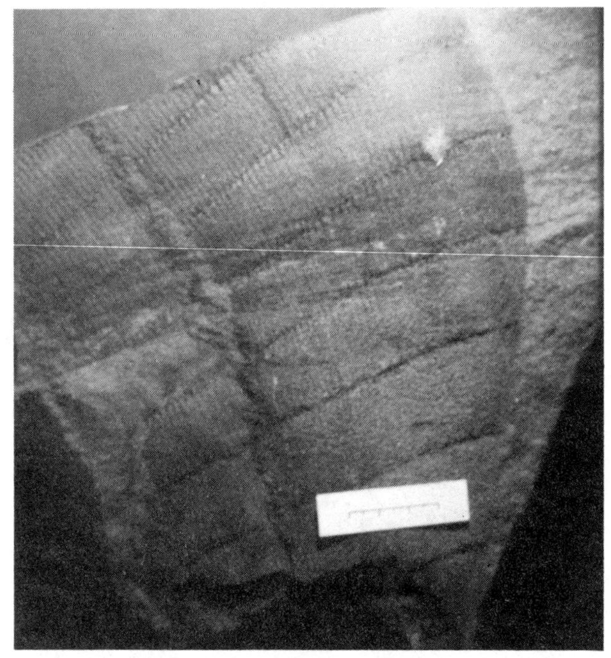

Fig. 1.8 — *Calamites* sp. Pith cast of a large sphenophyte. National Museum of Wales, Specimen No. 22.558G.2. From Pentwyn Quarry, South Wales, UK (Westphalian D).

1.5.1 Palaeoecology

Traditionally, this was taken to be the study of the interaction of past life with its environment. At least in palaeobotany, however, the topic now encompasses a much wider spectrum of study, covering all aspects of how a plant fossil is preserved. This includes the original controls on the growth and distribution of the plants (plant ecology), the processes by which plant fragments were transferred to the sediment in which they were to become entombed (taphonomy), and the various changes which occurred after burial (diagenesis). When treated in this broad sense, it clearly underpins most of the geological applications of plant fossils, which depends on their distribution pattern.

At a very simple level, plant fossils can provide an environmental message: well preserved, articulated plant fossils normally indicate non-marine or littoral deposits. At a more refined level, however, there are problems, particularly in the Palaeozoic. For one reason, the fossils are the remains of groups of plants with no, or only very distant, extant descendants, and it is thus impossible to have prior knowledge of their favoured habitats. Even where there are moderately close relatives living today, such as the sphenophytes (horsetails), it is far from certain that the Palaeozoic species occupied similar ecological niches to their living counterparts. The only answer is to start from scratch at well exposed localities, to integrate palaeobotanical and sedimentological observations, to try to build up an environmental model (examples of such work are reviewed in sections 2.4–2.9). The model can then be used to draw palaeoecological inferences from plant fossil distributions in less well exposed situations. Even here, though, great care has to be taken that such an extrapolation is valid and is not based on too generalized an interpretation of the original model (see section 2.10.5 for a further discussion on this point).

A further complication is that most plant fossils are not preserved in sediments representing the habitat of the original plant. Early studies on Palaeozoic plant palaeoecology tended to give little emphasis to this factor, and the results were thus often misleading. More recently, however, palaeoecologists have realized that *in situ* (autochthonous) fossils such as tree stumps give a far more reliable indication of the habitats favoured by these plants. A case in this author's particular field of interest concerns the diverse assemblages of pteridosperm and fern foliage found in the Upper Carboniferous flood-plain deposits of Britain; they have been interpreted as meaning that the flood-plains were occupied by ferns and pteridosperms (e.g. Scott 1979). However, it now transpires from work on *in situ* fossils (e.g. Gastaldo 1987) that these plants occupied the drier, raised levee banks and their foliage was transported into the flood-plains, probably during storm events; the flood-plains themselves, when not being flooded by sediment, were mainly occupied by lycophyte (club-moss) trees (see Cleal 1987; also section 2.8.1).

From the difficulties outlined above (and they are far from the only ones), it may be thought that it would be nigh-on impossible to use plant fossils as accurate palaeoenvironmental indicators. However, there has been considerable progress in this field, and it is becoming an increasingly useful geological tool. The way forward has proved to be the integration of primary geological (mainly sedimentological) and palaeobotanical data, to establish a model that can then provide information of interest to both the palaeobotanist (e.g. the habitat of the extinct plant) and the geologist (e.g. the depositional environment of the sediment).

1.5.2 Palaeogeography

Plants are essentially immobile organisms, whose migration is limited mainly to the dissemination of their propagules (seeds, spores, etc.). They thus tend to have more restricted geographical ranges than animals. Not surprisingly, therefore, plant fossils have been extensively used in palaeogeographical work, and they played a key role in the development of Wegener's continental drift model (reviewed by Bishop 1981).

The prime goal is to try to recognize particular geographical areas by natural associations of plant fossils. Such associations may be distinguished by endemic taxa or by a characteristic balance of taxa. The former are the easiest to recognize in practice, since they depend simply on the presence or absence of the characterizing taxa. The latter, in contrast, can only normally be recognized using statistical techniques such as cluster analysis or polar ordination (e.g. Raymond *et al.* 1985). Recent phytogeographical associations are established mainly using genera (e.g. Good 1974). Similarly, the genus/form-genus has generally been found to be the most useful for recognizing fossil associations (e.g. Chaloner & Lacey 1973), but for different reasons. Higher-ranking fossil taxa tend to be poorly defined and circumscribed, particularly when based on adpressions, whilst species and sub-specific taxa are too vulnerable to systematic subjectivity to be useful.

In Recent phytogeography, the plant associations are classified according to a nested hierarchy of units (termed phytochoria by Meyen 1987) reflecting different levels of similarity in composition. There is no unanimity as to the nomenclature for the different ranks of phytochoria (Allen & Dineley 1988) but perhaps the most widely used in work on the fossil record is that of Chaloner & Meyen (1973) and Meyen (1987). This recognizes four ranks of phytochoria — in descending order kingdom, area, province and district. These terms have been adapted from Recent phytogeographical work (e.g. Good 1974) and imply that there is an essential homology between the two types of classification. Raymond *et al.* (1985) and Allen & Dineley (1988) have pointed out, however, that the plant associations determined from the fossil record are quite different from Recent phytochoria. In addition to the potential distortion introduced by taphonomy, the fossil record tends only to sample a very limited range of lowland habitats, whereas Recent phytochoria should reflect the taxonomic composition of all habitats. Furthermore, fossil taxa tend to be based on vegetative morphology and may thus reflect ecophenotypic variation more than the phylogenetic position of the original plants.

Raymond *et al.* (1985) attempted to overcome this problem by referring to the assemblages of plant fossils simply as 'phytogeographical units'. However, this makes it difficult to establish a fully hierarchical classification, as is available with Meyen's system, without adopting a cumbersome nomenclature involving 'phytogeographical subunits' and 'phytogeographical subsubunits'. It is suggested instead, therefore, that the Meyen scheme be modified by prefixing his terms with 'palaeo-' (palaeokingdom, palaeoprovince, etc.), to indicate that they are based on the fossil record and are not fully homologous with Recent phytochoria. It must be emphasized that a palaeokingdom is not simply an ancient kingdom, as would be recognized by botanists living in (say) the Permian when interpreting their contemporary floras, but is a palaeophytogeographical unit established through the partial sampling provided by the fossil record. The difference is not merely semantic, but has potentially important consequences when trying to interpret the geographical distribution of plants from the fossil record.

Because of the fundamental differences between the phytogeography of living plants and the analysis of geographical distributions of plant fossils, the term *palaeophytogeography* has been used for the latter. It is unfortunately a rather cumbersome word. It seems better, however, to use such a compound word, whose roots will be immediately familiar to the scientists involved, than to follow current fashion and to coin another piece of alien jargon.

1.5.3 Biostratigraphy

As with other groups of fossils, the distribution of plant remains can be used to establish stratigraphical correlations. Such work is known as biostratigraphy. Plant biostratigraphy might be thought to relate to the genetic evolution of the parent plants, and there must of course be some underlying correlation between the two. However, the stratigraphical distribution of plant fossils is also controlled by a number of other and usually more potent factors. In essence, the lowest occurrence of a species of plant fossil in a stratigraphical sequence reflects the time when the parent plant and the environment of one of the lowland habitats became compatible. This may have been due to one of a number of causes.

(a) A species, through genetic drift, became adapted to the habitat.

(b) An environmental change caused the habitat to become suited to the species. This may be abiotic (e.g. a change in substrate conditions) or biotic (e.g. the introduction of a new animal species that provides a suitable vector for propagation).
(c) The extinction of another species freed a suitable niche in the habitat.
(d) The species migrated into the habitat from the one where it originally evolved (Laveine *et al.* 1989).

The last occurrence of a species is also controlled by a complex set of factors, although the most important are probably environmental shifts (similar to (b) above) or the introduction of a new and more successful plant species into the environment.

In the past, there was a tendency to make comments like '... the presence of this fossil assemblage shows that the strata are Upper Visean ...' or '... the junction between the Carboniferous and Permian in this section is defined by the lowest occurrence of this species ...'. However, such procedures are now recognized to be unsatisfactory and, in particular, prone to introducing circular argument. Consequently, what has become known as the biozonal method was introduced. A sequence of biozones are defined by the ranges of one or more fossil taxa, and attempts are made to locate the sequence in different stratigraphical sections. The identification of the boundary between biozones in different sections is then used as an indication of possible time correlation.

A number of different types of biozone are now recognized, depending on how their limits are defined (Bassett 1990). Almost without exception, however, the zones established for plant fossils are of the type known as *contiguous assemblage biozones*. Such zones are not defined by the range of a particular species. Instead, the limits of the zone are defined by the tops and/or bottoms of the ranges of several species. This has advantages over zones defined by single taxa (e.g. *local range biozones*), in which the absence of the index taxon can make it difficult to identify the biozone. That there is no identifying taxon or taxa, however, means that it can be difficult to position an isolated assemblage. Assemblage biozones are best applied to a sequence of assemblages, allowing the biozonal boundaries to be identified.

It is always possible that biozonal boundaries may not be isochronous, owing to migration of the taxa or differences in the timing of ecological changes. One way to check this is to identify the biozones in different areas where there is some independent, preferably abiotic, means of correlation (the tonsteins and marine bands of the Upper Carboniferous coalfields of Europe provide an excellent 'laboratory' for such work). If this is not available, then the only recourse is to make a detailed comparison of the stratigraphic ranges of different taxa in different areas, to see if the same pattern of tops and bottoms of ranges can be recognized. Since it is unlikely (but not impossible) that different species will be migrating and/or responding to ecological changes in the same way, a uniformity in pattern of range limits (i.e. tops and bottoms) will tend to confirm the robustness of the biostratigraphy for making time correlations.

There has been very little work on checking the robustness of plant biostratigraphies in the Palaeozoic, and there is still a tendency to take the results at face value. The most notable exceptions have been on the Upper Carboniferous coalfields of

Europe, mainly through the work of J.-P. Laveine (e.g. Laveine 1977, Josten & Laveine 1984, Cleal 1984, Zodrow & Cleal 1985). The results have on the whole been comfortingly consistent; at least within particular palaeoareas, the first occurrence of a particular species seems to be more or less isochronous (Laveine *et al.* 1989). However, great care has to be taken when attempting long-distance correlations, particularly between palaeoareas.

Finally, two alternative approaches to plant biostratigraphy should be mentioned in brief. First, there is the recognition of biozones on levels of evolutionary development. This has been most successfully developed by Banks in Dilcher & Taylor (1980), in a plant biostratigraphy for the Silurian and Devonian (although it should be pointed out that his biozones were also partly defined on the lines of traditional assemblage biozones). Such schemes are of great interest for outlining the broad trends of evolution in the plant kingdom. Because they are so loosely defined, however, biozones of this type will probably only be of limited use for practical geological purposes.

Secondly, there is what has become known as ecostratigraphy. This once-popular concept has tended to mean different things to different people. For some, it just means the introduction of ecological data into the biostratigraphic model, whereby meaning can be given to variations in fossil distribution, and thus to biozonal variations. This is obviously something to be recommended, at least in principle. For others (e.g. Krassilov 1978), however, it means the establishment of stratigraphic correlations on the basis of changes in organic communities. The idea is that such community changes tend to be triggered by major climatic or tectonic events, and should thus be recognizable over much greater geographical distances than the traditional biozones, and less vulnerable to the vicissitudes of local ecological variations. It is doubtful, however, if such 'ecozones' could ever produce the sort of resolution provided by traditional biozones. Furthermore, it is far from certain that ecological events that are supposed to alter the communities would influence different parts of the world at the same time. Nevertheless, this type of stratigraphy may have a role in establishing at least approximate correlations between palaeokingdoms where there are few if any taxa in common.

1.5.4 Palaeoclimatology

Palaeoclimatology is becoming an increasingly important factor in many branches of geological work. Climate can clearly affect things like sedimentation processes, erosion, weathering, and the build-up of economically important mineral deposits (e.g. bauxite). It hardly needs saying that there is a close relationship between climate and Recent plants and vegetational patterns. It might be expected, therefore, that Palaeozoic plant fossils would produce a useful palaeoclimatic signal. There are, however, major problems in interpreting that signal. Because Palaeozoic and Recent plants are so distantly related to each other, it is far from certain that they will react in the same way to climatic influences. Nevertheless, there are some aspects of Palaeozoic plant fossils that may provide a useful indicator of climatic variations.

There are two basic ways of looking at plants from a climatic standpoint. Firstly, there is *vegetational physiognomy*. This is where plants develop certain specific morphological or anatomical characters in response to climatic factors. Secondly, there is the recognition of general changes in vegetational communities, which may have been triggered by climate. It is not always clear what sort of climatic change (e.g. increasing aridity, decreasing temperature) is being reflected by a particular vegetational change in the Palaeozoic, although this may become evident on the basis of vegetational physiognomy.

1.6 CHRONOSTRATIGRAPHICAL NOMENCLATURE

The chronostratigraphy used in this book essentially follows that recently summarized by the IUGS Commission on Stratigraphy (Cowie & Bassett 1989). However, there are a few minor differences and points that require clarification.

The lower two stages of the Devonian have recently been changed from Gedinnian and Siegenian to Lochkovian and Pragian, respectively (Ziegler & Klapper 1985). Not only were the names changed, but there were also minor adjustments to the positions of the stage boundaries. As pointed out by Edwards (1990), this may provide better refinement in marine sequences, but difficulties of detailed correlation make it impossible to identify the new boundaries accurately in non-marine sequences. Consequently, the Gedinnian and Siegenian stages traditionally used in palaeobotanical work (at least, over the last 30 years) have been retained in this volume.

It has long been accepted that the Carboniferous has to be divided into subdivisions, but until recently there has been little agreement as to the number and definitions of the subdivisions. However, following intense international collaboration, it is now generally accepted that there will be two subsystems, and that the resulting mid-Carboniferous boundary will be identified by conodont criteria (Lane *et al.* 1985). The boundary coincides approximately with the Mississippian-Pennsylvanian boundary in North America, the Arnsbergian-Chokierian boundary in Europe, and the Sepukhovian-Bashkirian boundary in the USSR. Throughout this volume, the terms Upper Carboniferous and Lower Carboniferous have been used in this context, and not in the traditional European sense (i.e. Dinantian-Silesian), or in that traditionally used by Soviet stratigraphers.

Through much of the palaeoequatorial belt, the so-called Heerlen Classification has become the chronostratigraphy normally used in the Upper Carboniferous (reviewed by Wagner 1974). Since it was first developed during the 1927 and 1935 Carboniferous congresses at Heerlen, the scheme has undergone a number of changes, most significantly in that the three major divisions of the Upper Carboniferous (Namurian, Westphalian, Stephanian) have been raised in rank from stage to series. This resulted in changes in nomenclature for the subdivisions of these units; names such as Namurian A and Westphalian C were regarded as unsuitable for stages. Most of these nomenclatural changes are discussed in Wagner's paper, but some have been introduced since 1974. These are as follows.

Old Name/New Name	
Stephanian A	= Barruelian
Westphalian C	= Bolsovian
Westphalian B	= Duckmantian
Westphalian A	= Langsettian

The positions of these stages within the rest of the Heerlen Classification is shown in Fig. 4.1 (N.B. formal names have still to be proposed for the Westphalian D, Stephanian B and Stephanian C).

There is still no consensus as to the position of the Carboniferous–Permian boundary. When agreement is finally reached, it will no doubt be defined in marine sequences, and it will be extremely difficult to locate in the type of non-marine sequence in which most plant fossils occur (Kozur 1984). In the classic Russian sections, the boundary is placed at the base of the Asselian Stage identified at the base of the *Schwagerina vulgaris–S. fusiformis* fusilinid zone (Rauser-Chernousova & Shchegolev in Meyen 1980) and this is also used in the Angaran areas (e.g. Gorelova *et al.* 1973). In Laurasia, the junction is taken at the Stephanian C–Autunian boundary. The boundary in Gondwana has traditionally been linked with the base of the *Glossopteris* Biozone, although it is now recognized that some glossopterids probably appeared below the Permian in the classic Russian sense (Roberts in Wagner *et al.* 1983). Rocha Campos & Archangelsky in Wagner *et al.* (1983) use radiometric evidence to try to produce a more refined boundary in Gondwana, but the results are far from convincing. In Cathaysia, the position is even more confusing, the boundary being defined in south China by the top of the *Pseudoschwagerina* fusilinid biozone, and in north China on rather imprecise palaeobotanical arguments (Yang, Li & Gao in Wagner *et al.* 1983).

In view of the problems in defining and recognizing the Carboniferous–Permian boundary in a global context, the locally defined boundaries have been used in this volume, and there is no implication that they are exactly coincident. This should be borne in mind in any discussions on global changes occurring at or about this boundary.

REFERENCES

Allen, K. C. & Dineley, D. L. (1988) Mid-Devonian to mid-Permian floral and faunal regions and provinces. In: Harris, A. L. & Fettes, D. J. (eds) *The Caledonian–Appalachian Orogeny. Geol. Soc. London, Spec. Publ.* **38** 531–548.

Banks, H. P. (1972) *Evolution and plants of the past*. Macmillan, London.

Barthel, M. (1968) *'Pecopteris' feminaeformis* (Schlotheim) Sterzel und *'Araucarites' spiciformis* Andrae in Germar — Coenopteriden des Stefans und Unteren Perm. *Paläontol. Abh. Abt. B* **2** 727–742.

Bassett, M. G. (1990) Zone fossils. In: Briggs, D. E. G. & Crowther, P. R. (eds) *Palaeobiology. A synthesis*. Blackwell Scientific Publications, Oxford, pp. 466–467.

Bishop, A. C. (1981) The development of the concept of continental drift. In: Cocks, L. R. M. (ed.) *The evolving earth*. British Museum (Natural History), London, and The University Press, Cambridge, pp. 155–164.

Chaloner, W. G. & Lacey, W. S. (1973) The distribution of Late Palaeozoic floras. In: Hughes, N. F. (ed.) *Organisms and continents through time. Spec. Pap. Palaeont.* **12** 271–289.

Chaloner, W. G. & Meyen, S. V. (1973) Carboniferous and Permian floras of the northern continents. In: Hallam, A. (ed.) *Atlas of palaeobiogeography*. Elsevier, Amsterdam, pp. 169–186.

Cleal, C. J. (1984) The Westphalian D floral biostratigraphy of Saarland (Fed. Rep. Germany) and a comparison with that of South Wales. *Geol. J.* **19** 327–351.

Cleal, C. J. (1986) Identifying plant fragments. In: Spicer, R. A. & Thomas, B. A. (eds) *Systematic and taxonomic approaches in palaeobotany. Systematics Association Special Volume 31*. University Press, Oxford, pp. 53–65.

Cleal, C. J. (1987) This is the forest primaeval. *Nature* **326** 828.

Cleal, C. J. (1988) British palaeobotanical sites. *Spec. Pap. Palaeontol.* **40** 57–71.

Cleal, C. J. & Zodrow, E. L. (1989) Epidermal structure of some medullosan *Neuropteris* foliage from the middle and upper Carboniferous of Canada and Germany. *Palaeontology* **32** 837–882.

Collinson, M. E. (1987) Special problems in the conservation of palaeobotanical material. *Geol. Curator* **4** 439–445.

Cowie, J. W. & Bassett, M. G. (1989) International Union of Geological Sciences 1989 *Global Stratigraphic Chart. Episodes 12* (Supplement).

Dilcher, D. L. & Taylor, T. N. (eds) (1980) *Biostratigraphy of fossil plants. Successional and paleoecological analyses*. Dowden, Hutchinson & Ross, Stroudsburg.

Ferguson, D. K. (1985) The origin of leaf assemblages — new light on an old problem. *Rev. Palaeobot. Palynol.* **46** 117–188.

Flügel, E. (ed.) (1977) *Fossil algae. Recent results and developments*. Springer, Berlin.

Galtier, J. (1986) Taxonomic problems due to preservation: comparing compression and permineralized taxa. In: Spicer, R. A. & Thomas, B. A. (eds) *Systematic and taxonomic approaches in palaeobotany. Systematics Association Special Volume 31*. University Press, Oxford, pp. 1–16.

Gastaldo, R. A. (1987) Confirmation of Carboniferous clastic swamp communities. *Nature* **326** 869–871.

Good, R. (1974) *The geography of flowering plants*. 4th edn. Longman, London.

Gorelova, S. G., Men'shikova, L. V. & Khalfin, L. L. (1973) Fitostratigrafiya i opredelitel' ractenii verkhnepaleozoiskikh uglenosnykh otlozhenii Kuznetskogo Besseina. *Trudy S.N.I.I.G.G.I.M.Ca.* **140** 1–169 [In Russian].

Halle, T. G. (1933) The structure of certain fossil spore-bearing organs believed to belong to pteridosperms. *K. Svensk. Vetenskapsakad. Hand.* **12** 1–103.

Hughes, N. F. (1989) *Fossils as information*. University Press, Cambridge.

Josten, K.-H. & Laveine, J.-P. (1984) Paläobotanisch-stratigraphische Untersuchungen im Westfal C-D von Nordfrankreich und Nordwestdeutschland. *Fortschr. Geol. Rheinld u. Westf.* **32** 89–117.

Joy, K. W., Willis, A. J. & Lacey, W. S. (1956) A rapid cellulose peel technique in palaeobotany. *Ann. Bot. (New Series)* **20** 635–637.

Kerp, J. H. F. (1988) Aspects of Permian palaeobotany and palynology. X. The west- and central-European species of the genus *Autunia* Krasser emend. Kerp (Peltaspermaceae) and the form-genus *Rhachiphyllum* Kerp (callipterid foliage). *Rev. Palaeobot. Palynol.* **54** 135–150.

Kozur, H. (1984) Carboniferous–Permian boundary in marine and continental sediments. *C. R. 9e Congr. Int. Strat. Géol. Carbon.* (Washington & Urbana, 1979) **2** 577–586.

Krassilov, V. A. (1975) *Paleoecology of terrestrial plants: basic principles and techniques*. Wiley, New York.

Krassilov, V. A. (1978) Organic evolution and natural stratigraphical classification. *Lethaia* **11** 93–104.

Lane, H. R., Bouckaert, J., Brenckle, P. L., Einor, O. L., Havlena, V., Higgins, A. C., Yang J. Z., Manger, W. L., Nassichuk, W., Nemirovskaya, T., Owens, B., Rammsbottom, W. H. C., Reitlinger, E. A. & Weyant, M. (1985) Proposal for an international Mid-Carboniferous boundary. *C. R. 10e Congr. Int. Strat. Géol. Carbon. (Madrid, 1983)* **4** 323–339.

Laveine, J.-P. (1977) Report on the Westphalian D. In: Holub, V. M. & Wagner, R. H. (eds) *Symposium on Carboniferous Stratigraphy. Geological Survey, Prague*, pp. 71–81.

Laveine, J.-P., Zhang S. & Lemoigne, Y. (1989) Global paleobotany, as exemplified by some Upper Carboniferous pteridosperms. *Bull. Soc. Belge Géol.* **98** 115–125.

Leclercq, S. (1960) Refendage d'une roche fossilifère et dégagement de ses fossiles sous binoculaire. *Senckenberg. Leth.* **41** 483–487.

Meyen, S.V. (1980) *Biostratigrafiya pogranichnykh otlozhenii Karbona i Permi*. Order of the Red Banner of Labour Geological Institute, Moscow [In Russian].

Meyen, S. V. (1987) *Fundamentals of palaeobotany*. Chapman & Hall, London.

Mosbrugger, V. (1983) Organische Zusammengehörigkeit zweier Fossil-Taxa als taxonomisches Problem am Beispiel der jungpaläozoischen Farnfruktifikationen *Scolecopteris* und *Acitheca*. *Rev. Palaeobot. Palynol* **40** 191–206.

Poort, R. J. & Kerp, J. H. F. (1990) Aspects of Permian palaeobotany and palynology. XI. On the recognition of true peltasperms in the Upper Permian of western and central Europe and a reclassification of species formerly included in *Peltaspermum* Harris. *Rev. Palaeobot. Palynol.* **63** 197–225.

Raymond, A., Parker, W. C. & Parrish, J. T. (1985) Phytogeography and paleoclimate of the Early Carboniferous. In: Tiffney, B. H. (ed.) *Geological factors and the evolution of plants*. University Press, Yale, pp. 169–222.

Schaarschmidt, F. (1975) Farbaufnahmen von Pflanzenfossilien in polarisiertem Licht. *Cour. Forschungs Inst. Senckenberg* **13** 145–148.

Scott, A. C. (1979) The ecology of Coal Measure floras from northern Britain. *Proc. Geol. Ass.* **90** 97–116.

Scott, A. C. & Collinson, M. E. (1983) Investigating fossil plant beds. Part 2: Methods of palaeoenvironmental analysis and modelling and suggestions for experimental work. *Geology Teaching* **8** 12–26.

Shute, C. H. & Cleal, C. J. (1987) Palaeobotany in museums. *Geol. Curator* **4** 553–559.

Stewart, W. N. (1983) *Paleobotany and the evolution of plants*. University Press, Cambridge.

Taylor, T. N. (1981) *Paleobotany. An introduction to fossil plant biology*. McGraw-Hill, New York.

Thomas, B. A. & Spicer, R. A. (1986) *The evolution and palaeobiology of land plants*. Croom Helm, London.

Traverse, A. (1988) *Paleopalynology*. Unwin Hyman, London.

Tucker, M. E. (1982) *The field description of sedimentary rocks*. The Geological Society, London.

Vakhrameev, V. A., Dobruskina, I. A., Meyen, S. V. & Zaklinskaja, E. D. (1978) *Paläozoische und mesozoische Floren Eurasiens und die Phytogeographie dieser Zeit*. Gustav Fischer, Jena.

Visscher, H., Kerp, J. H. F. & Clement-Westerhof, J. A. (1986) Aspects of Permian palaeobotany. VI. Towards a flexible system of naming Palaeozoic conifers. *Acta Bot. Neerl.* **35** 87–99.

Wagner, R. H. (1974) The chronostratigraphic units of the Upper Carboniferous in Europe. *Bull. Soc. Belge Géol.* **83** 235–253.

Wagner, R. H., Winkler Prins, C. F. & Granados, L. F. (eds) (1983) *The Carboniferous of the world. I. China, Korea, Japan & S. E. Asia*. Inst. Geol. y Min. de España, Madrid (I.U.G.S. Publication No. 16).

Wagner, R. H., Winkler Prins, C. F. & Granados, L. F. (eds) (1985) *The Carboniferous of the world. II. Australia, Indian Subcontinent, South Africa, South America, & North Africa*. Inst. Geol. y Min. de España, Madrid (IUGS Publication No. 20).

Ziegler, W. & Klapper, G. (1985) Stages of the Devonian System. *Episodes* **8** 104–109.

Zodrow, E. L. & Cleal, C. J. (1985) Phyto- and chronostratigraphical correlations between the late Pennsylvanian Morien Group (Sydney, Nova Scotia) and the Silesian Pennant Measures (south Wales). *Can. J. Earth Sci.* **22** 1465–1473.

2

Palaeoecology

R. M. Bateman

Although definitions of ecology differ widely, not even the most generalized (e.g. 'the totality or pattern of relationships between organisms and their environment': Odum 1971: 3) approaches the scope traditionally encompassed by palaeoecology. For example, Meyen (1987: 271–283) and Krassilov (1975) listed preservational modes, taphonomy, and plant reconstruction as aspects of palaeoecology. These phenomena are inconsequential for neoecological studies of living organisms but profoundly influence palaeoecological interpretation, and are therefore subject to detailed discussion in this chapter. Fig. 2.1 shows their relationships to other palaeobotanical disciplines. A necessarily brief synthesis of current theories on the evolution of land plants and land plant communities is followed by several case studies that integrate palaeobiological and geological knowledge. They are selected from the now substantial literature on the Palaeozoic palaeoecology of the best documented phytochorion, Eurameria. They illustrate the strengths and weaknesses of different approaches to different types of fossil plant assemblages from different time intervals. The paper concludes with highly personal accounts of some crucial topical controversies in Palaeozoic palaeoecology.

I. Beautiful theories

2.1 DIAGENESIS AND PRESERVATION

Animals are more commonly represented in the fossil record than plants. This largely reflects the more frequent occurrence in many faunal clades of species characterized by extensive hard parts, resistant to both compression and oxidation (duripartic preservation *sensu* J. M. Schopf 1975), and the dominance of marine over non-marine deposition. Although the inferior preservation potential of plants has

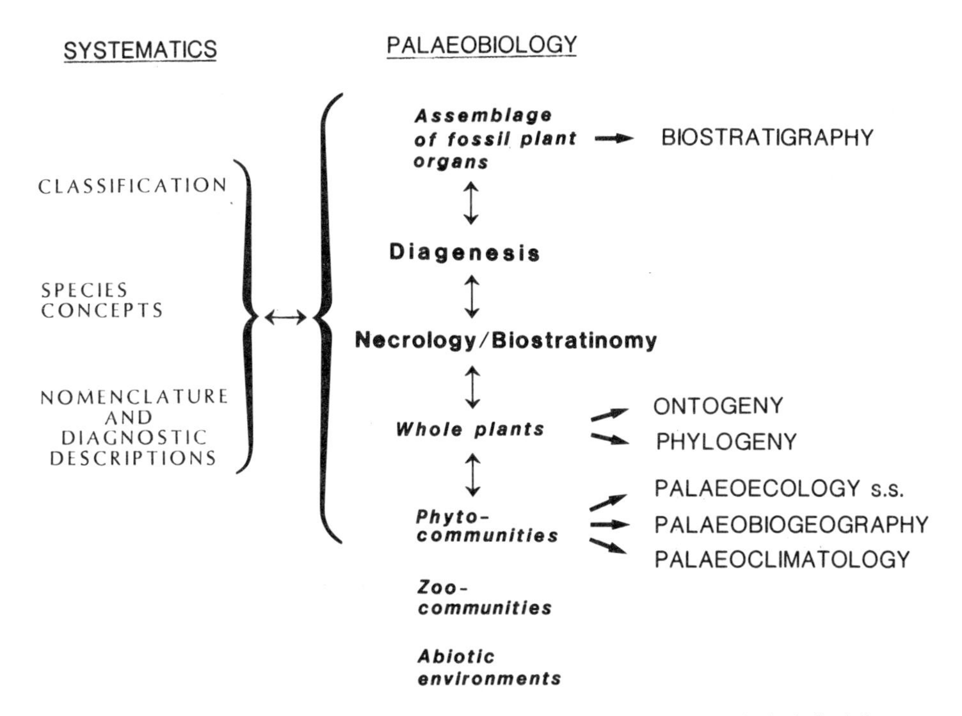

Fig. 2.1 — Conceptual relationship of palaeoecology to other palaeontological disciplines. Biological entities are italicized and categories of derived information are capitalized. Diagenesis and taphonomy are sets of inter-related, environmentally determined processes.

often been exaggerated, the paucity of mineralized hard parts in most vascular plants and their confinement during growth to the terrestrial realm (which experiences net denudation) inevitably result in a record that is localized in both time and space.

2.1.1 Tissue stability

The most serious challenge to plant preservation is oxidative biodegradation. The only duripartic hard parts commonly found in land plants are opaline silica bodies termed phytoliths. These dominantly intracellular mineral concentrations are abundant in extant descendants of many Palaeozoic lineages (e.g. *Selaginella*, *Equisetum*, some hymenophyllacean and polypodiacean ferns: Piperno 1988). Regrettably, they have been wholly ignored by Palaeozoic palaeobotanists. Of the better-known but less-durable amineralized tissues, cell contents (protoplasm and organelles) generally decay before the cell wall, whatever its composition. Variation in wall composition determines the relative susceptibilities of different plant materials to degradation:

phloem (1)>internal parenchyma (2)>epidermal parenchyma(3)>
sclerenchyma (4)>xylem (5)>cuticle (6)>spores/pollen (7)>phytoliths (8).

The loss of each cell/tissue type can be regarded as a degradational stage (e.g. loss of phloem is Stage 1). The precise order of degradation can deviate from this stability series, depending on the nature of the plant, its environment, and the relevant degradative processes. The relative resistancess of cell types (1)–(5) largely reflect the relative proportion of the polysaccharide cellulose to the phenyl polymer lignin in their cell walls (Rolfe & Brett 1969). Cuticle (6), which envelops all epidermal tissue, is a mixture of the lipid polymer cutin and various waxes. Land plant spores *s.l.* (including pollen grains: 7) are enclosed by exines composed of the carotenoid polymer sporopollenin. The comparative resistance of cuticle and spores to post-mortem decay reflects adaptation to pre-mortem decay; it is essential that plant tissues exposed to the external environment, notably the epidermis and various types of disseminule (see below), should resist desiccation and biodegradation while alive. Their persistence after death is of no advantage to the plant but of considerable advantage to the palaeobotanist.

2.1.2 Diagenesis

To allow any form of plant preservation, the depositional environment must inhibit biodegradative bacteria and fungi; even then, autolytic degradational processes within the pre-fossil (i.e. the self-digestion of cells by their own enzymes and an inevitable consequence of the disruption of biochemical cycles upon death) are unlikely to be suppressed. Post-burial modification of the prospective fossil is intimately bound with lithification of the enclosing sediment, notably reduction in pore space (and consequently in interstitial fluids and gases). This can occur physically, by compression of the sediment due to overburden pressure, and chemically, by diagenetic mineralization. The relative speed and severity of degradation, compression and mineralization determine whether the potential plant fossil will be preserved and in what form.

After the rapid (dominantly autolytic) decay of the protoplasm, the volume of the buried plant consists largely of intracellular spaces (Fig. 2.2(a). Sooner or later, the fossil is subject to sufficient overburden pressure to eliminate these spaces by compression perpendicular to the bedding, reducing the third dimension of the fossil and thereby erasing most anatomical details (Fig. 2.2(b)). Three-dimensionality can only be maintained if the resistance to compression of the fossil (and usually also of the surrounding matrix) is increased by mineralization. Either intracellular spaces become infilled with minerals precipitated from pore fluids (permineralization *sensu* Schopf 1971, 1975; Fig. 2.2(c), (d)) or the entire fossil becomes entombed in diagenetic minerals (authigenic cementation: Fig. 2.2(e)). Permineralization can provide an effective seal against further decay of the organic matter (Fig. 2.2(c)); even if the cell walls are subsequently replaced during later phases of permineralization, cell boundary information often persists in the resulting inorganic pseudomorph (Fig. 2.2(d)). Authigenic cementation is less intimate and pervasive, usually allowing loss of organic matter and preserving only gross morphology as mould and cast replicas (Fig. 2.2(e)). Thus, maintenance of three-dimensionality does not necessarily result in retention of anatomical detail.

2.1.3 Classification of preservation states

Unfortunately, as noted (and practised) by Schopf (1975: 49), attempts to classify plant fossils according to their preservation states have repeatedly (and inconsistently) confused process and product in their terminology. Fig. 2.3 places the most

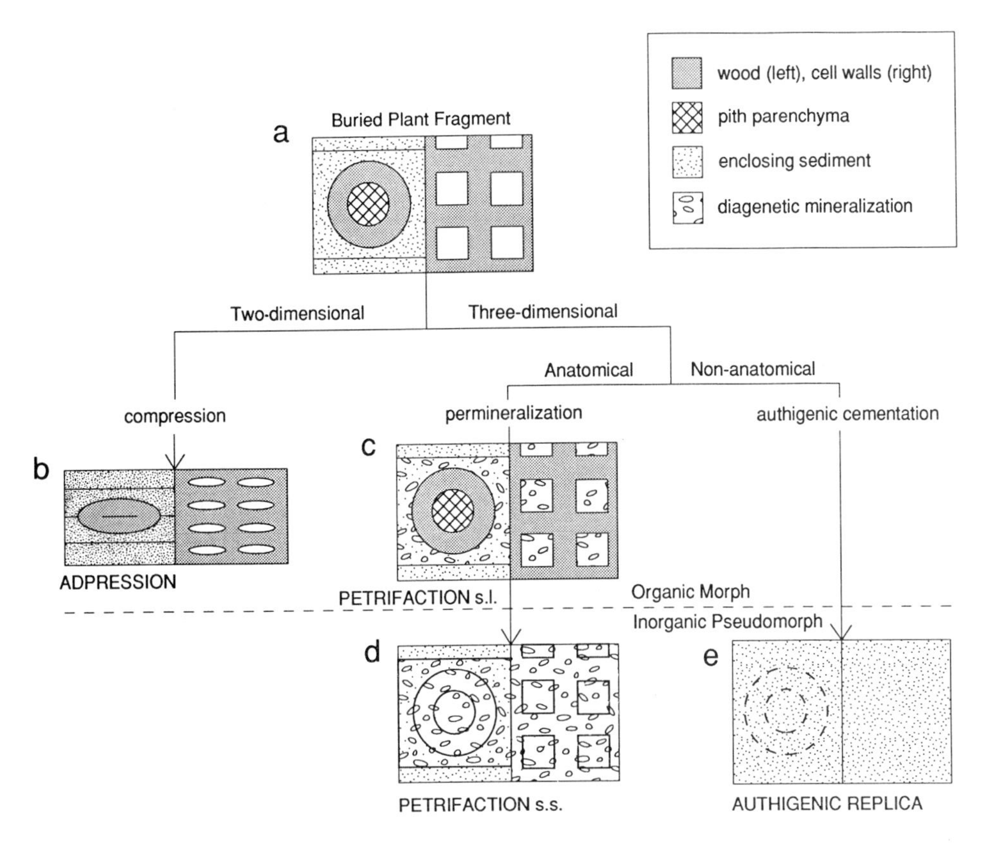

Fig. 2.2 — Summary of modes of fossil plant preservation. (Includes concepts from Scott & Collinson 1983, Figs 6–7, Thomas & Spicer 1987, Fig. 42.)

commonly encountered terms in the context of (1) progressive shape changes, resulting from physical compression, and (2) progressive compositional changes, resulting from loss of volatile components from the organic matter (devolatilization) and subsequent loss of the residual carbon (decarbonization), leaving only an inorganic pseudomorph. The chosen threshold between three-dimensional petrifactions (Gr. petra=rock, L. factus=make) and two-dimensional adpressions (L. adpressus=lie flat against: Shute & Cleal 1987: 559) is a reduction to 20% of the original thickness; this approximates the degree of compression required to eliminate intracellular spaces from most tissues. Devolatilization (coalification *sensu* Schopf 1975) alters cell walls from yellow-brown and plastic to black and brittle. Subsequent loss of coalified material (decarbonization) provides a less ambiguous preservational threshold from organic morph to inorganic pseudomorph. Thus, the composition of a plant fossil can be conveyed as an adjective and its physical state as a noun (e.g. devolatilized adpression, decarbonized petrifaction).

The relationships of these novel terms to more conventional descriptors of preservational modes (particularly those in Schopf's (1975) influential classification)

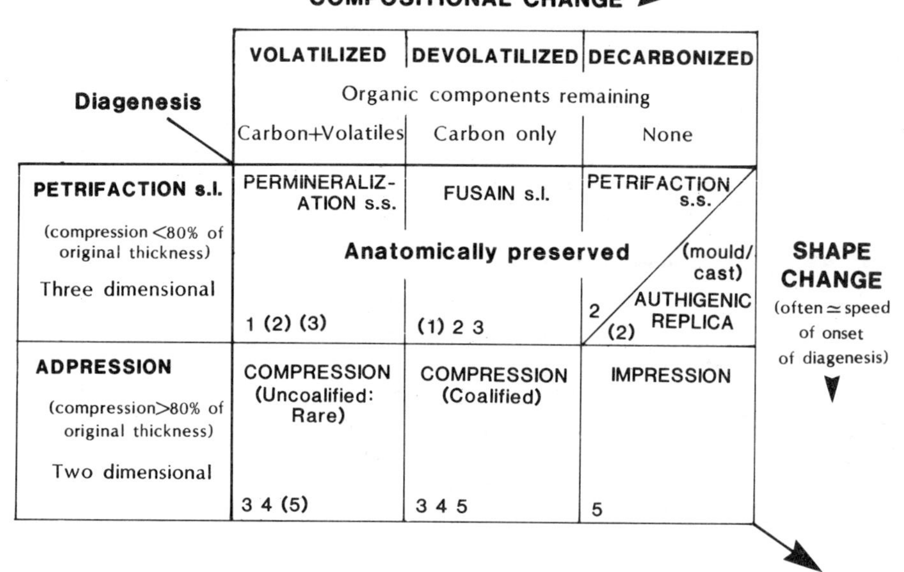

Fig. 2.3 — Comparison of novel classification of preservation states of plant fossils, based on diagenetic modification of shape and chemical composition, with the widely used classification of Schopf (1975). Numbers indicate preparation techniques (parenthetic techniques are less commonly applied to that preservation state) 1, acetate peels (Joy *et al.* 1956, Phillips *et al.* 1977); 2, petrographic thin sections (Murphy 1986), 3, maceration (Cutler *et al.* 1982, Traverse 1988); 4, resin transfer; 5, latex replicas (Chaloner & Gay 1973, Hill 1986). (See also Kummel & Raup 1965, Tiffney 1989.)

are shown in Fig. 2.3. Their significance is best illustrated using specific examples of preservational histories.

Initially, compression is accommodated by the intracellular spaces and the cell walls remain largely unaltered (Fig. 2.2(a)). However, compression usually continues unabated, driving off volatile components from the cell walls and progressively coalifying the adpression. Consequently, volatilized adpressions are unknown in the Palaeozoic and rare in younger strata. In contrast, devolatilized adpressions (coalified compressions *sensu* Schopf 1975) are the most common type of plant megafossil; the compressed coalified tissue has been termed a phytoleim (Kryshtofovich 1944). Although phytolemmas are generally internally amorphous, the relatively resistant cuticular envelope often persists, preserving a replica of the external cellular morphology of the epidermis. Occasionally, cuticles persist while the enclosed organic matter does not, and can become concentrated to form rare 'paper coals' (e.g. Neavel & Guennel 1960). Similarly, compressed sporangia often retain spores and compressed ovules retain megaspore membranes, albeit similarly two-dimensional.

In closed systems, decarbonization requires intense temperatures and pressures. Decarbonized adpressions (impressions *sensu* Schopf) more commonly result from 'opening' of fossiliferous deposits, particularly recent exposure to pedochemical weathering. This allows exfoliation of the phytoleim, leaving only a replica (cast) of the fossil impressed into the rock. Further impressions often result when a palaeobotanist cleaves a rock along a fossiliferous bedding plane; if the entire organic phytoleim adheres to the part, the counterpart bears only an inorganic impression. The part in turn becomes an impression if the highly friable phytoleim exfoliates during specimen preparation or in storage. Thus, adpression assemblages often consist of a mixture of coalified compressions and inorganic impressions. Such assemblages have traditionally been termed 'compression–impression' floras; the awkwardness of this term prompted its replacement with 'adpression' (Shute & Cleal 1987).

Terminology is even more problematic for fossils preserved in three dimensions (Fig. 2.3). The process of permineralization (well defined by Schopf 1975) has also been used as a noun to describe a mode of preservation, but the range of preservation states encompassed by that mode has not been carefully defined. Schopf (1975) employed the term for all anatomically preserved plant fossils, though some subsequent authors (e.g. Scott & Collinson 1983: 117, Thomas & Spicer 1987: 44) misrepresented Schopf as arguing that anatomically preserved fossils lacking organic matter (Fig. 2.2(d)) should be termed petrifactions (or 'petrifications'), and that only those fossils retaining organic matter (Fig. 2.2(c)) should be termed permineralizations. I have chosen to refer to all three-dimensionally preserved fossils (even those lacking anatomical detail) as petrifactions. The quality of anatomical preservation can be exceptional (e.g. Stewart 1983: 14–17).

Volatilized petrifactions retain brown cell walls and are exemplified by Upper Carboniferous coal balls (e.g. Scott & Rex 1985). The occurrence of devolatilized petrifactions (characterized by black, brittle, coalified cell walls) has rarely been acknowledged; they are unpopular with most palaeobotanists as they provide poor acetate peels, though they are readily thin-sectioned. There remains an ongoing controversy concerning the formation of three-dimensional coalified plant fossils (broadly termed fusain). Some authors (e.g. Harris 1981, Cope & Chaloner 1985, Scott 1989) argued that most such fossils are *char*coalified pyrofusain, formed by pyrolytic devolatilization as a result of exposure to wildfire prior to burial. Others (e.g. Schopf 1975, Beck *et al.* 1982) claim a predominance of diagenetically formed biofusain, formed by geothermal devolatilization long after burial. Less commonly, plant fragments are charred *during* burial, notably those interred during volcanic eruptions. In any of these cases, devolatilization renders the fossils relatively brittle and prone to disintegration during compaction of the enclosing sediment; most Palaeozoic fusain owes its continued existence to physical support afforded by diagenetic mineralization.

The most common permineralizing agents are carbonates (notably calcite, e.g. Upper Carboniferous coal balls: Scott & Rex 1985), silica (e.g. Lower Devonian Rhynie locality: Knoll 1985a), and pyrite (e.g. Lower Devonian Targrove Quarry: Kenrick & Edwards 1988), though phosphates and iron oxides/hydroxides are also important (Schopf 1975). Minerals carried in solution and precipitated in or on the plant fossil are often generated by weathering elsewhere in the succession, either as a result of subaerial pedogenesis or geochemical solution at depth. This requires a source of readily dissolved minerals (e.g. volcanic ashes and limestones). Alternatively, the permineralizing agent can precipitate biogenically rather than diagenetically. Examples include permineralization in bacteriogenic pyrite (e.g. Kenrick & Edwards 1988) and phosphatic permineralization in coprolites (e.g. Scott & Taylor 1983). Many agents often occur in a single permineralized fossil and several phases of permineralization are often evident in petrographic thin sections (e.g. Rex 1986b, Rex & Scott 1987, Bateman & Scott 1990); unfortunately, authors tend to report only the most abundant infilling mineral. Plant fragments often act as nuclei for diagenetic cementation of the surrounding sediment. In such cases, the early stages of cementation result in nodules whose shape is influenced by the enclosed plant fossil. Expansion of the nodules can result in coalescence (seen in many coal balls) and eventually in an extensive laminar concretionary horizon, particularly if the mineral-bearing fluids have pooled immediately above a less porous horizon.

If successive phases of permineralization are separated by a period of organic decay, the resulting decarbonized petrifaction retains at least some anatomical detail despite being an inorganic replica (Fig. 2.2(d)). Authigenic cementation, which can be assisted by mineralization prior to burial (Spicer 1977), rarely infiltrates intracellular spaces, usually preventing retention of anatomical detail (Fig. 2.2(e)). Decay of the plant fragment leaves a void that can persist but is usually infilled, typically syndepositionally with sediment (Rex 1985) rather than post-depositionally with precipitated minerals. This produces three-dimensional moulds and casts of the exterior of the specimen or, in some cases, of internal cavities prone to infilling. External moulds/casts are well illustrated by the iron-rich Mazon Creek nodules from the Upper Carboniferous of Illinois (e.g. Darrah 1969) and occurrences of *in situ* infillings of rootstocks and stem bases, particularly of arboreous lycopsids (Grierson & Banks 1963, MacGregor & Walton 1972). Well-known examples of internal moulds/casts include the widespread pithcasts of Upper Carboniferous calamites, and similarly infilled *in situ* root channels are the closest palaeobotanical analogue to faunal trace fossils. Small-scale moulds and casts also occur in permineralized specimens, such as internal casts of spore exines (Bateman in press).

Fossils also transgress boundaries between preservational categories as diagenesis proceeds. Even subtle differences in preservation can result in considerable differences in the appearance of specimens. Such preservation polymorphs (*sensu* Rowe 1988) are often mis-assigned to different genera, as exemplified by the different degrees of decortication and compression shown by various devolatilized adpressions of arboreous lycopsids. Studies of preservation polymorphs can provide information about the biology of the plants as well as their preservational histories (e.g. Rex & Chaloner 1983, Rex 1986a). Even single specimens can show a range of preservation states (e.g. Galtier 1986). Very localized nodular permineralization of specimens otherwise preserved as devolatilized adpressions has been especially important in providing information on Devonian (e.g. Kenrick & Edwards 1988) and Lower Carboniferous (e.g. Walton 1949, Jennings 1976) plant assemblages, allowing determination of both anatomical detail and gross morphology.

2.2 TAPHONOMY

2.2.1 Conceptual basis

The fossil record undoubtedly provides a heavily biased view of Palaeozoic biotas, particularly those living in the dominantly erosional terrestrial realm. Less certain are the overall degree of bias, the relative importance of various potential causes of bias, and the optimal techniques to correct for those biases. These concerns have fuelled the rise of the discipline of taphonomy: the study of 'the transition of organic remains from the biosphere to the lithosphere' (Efremov 1940). The importance of taphonomy stems from the fact that no biocommunity escapes post-mortem modification, and that the problem is especially acute with vascular plants. These sedentary, typically highly branched organisms almost inevitably experience disarticulation into component organs or organ aggregates (here termed 'plant parts'), as a result of both pre-mortem dehiscence/abscission and post-mortem disintegration. Many such plant parts are valuable biostratigraphical tools (Chapter 5). However, palaeobiological studies of ontogeny and phylogeny require reconstruction of the plants (Chaloner 1986, Bateman & Rothwell 1990), while satisfactory ecological, biogeographical (as opposed to palaeophytogeographical — see section 1.5.2) and palaeoclimatological interpretations of fossil floras also require assignment of reconstructed whole-plant species to their communities of origin (Fig. 2.1).

It has become fashionable to divide taphonomy into the broadly sequential subdisciplines of necrology (abscission, death, decay), biostratinomy (transport, deposition, burial), and diagenesis (post-burial physicochemical modification, lithification) (Behrensmeyer & Kidwell 1985, Gastaldo 1988). These three phases form the basis of the taphonomic cycle (Fig. 2.4). Biostratinomy and diagenesis constitute the classic sedimentary cycle of entrainment>transport>deposition>burial>post-burial modification. The periods spanned by each of these events can range from instantaneous (entrainment, deposition) through days or months (transport) to millennia (diagenetic mineralization), and final burial is often preceded by a series of transient 'metaburials' reflecting periods separating repeated sedimentary cycles. Modifying processes operating during such cycles obscure the provenance of both inorganic components studied by the sedimentologist and organic components studied by the palaeontologist. The most logical method of reconstructing provenance is to trace the sedimentary history of an assemblage of clasts back through the sedimentary cycle (and thereby through time), partitioning the cycle into stages and assessing the effects of modifying processes potentially operative at each stage (Bateman 1989). Although the main objectives of this approach are to reconstruct plant species and communities, it also yields much valuable information on environments of growth, transport and deposition.

Focusing primarily on organic rather than inorganic particles increases the range of modifying processes (taphonomic filters *sensu* Behrensmeyer & Kidwell 1985) that must be considered; classic sedimentological biases such as sorting and mechanical abrasion are joined by pyrolysis and biodegradation. Some modifying processes are confined to a single stage of the cycle (e.g. sorting and mechanical abrasion can only occur during transport) but others can operate during all necrological, biostratinomic and diagenetic phases (e.g. herbivory, microbial decay, autolysis: Fig. 2.4). Taphonomic processes also tend to be interdependent. For example, burning rapidly

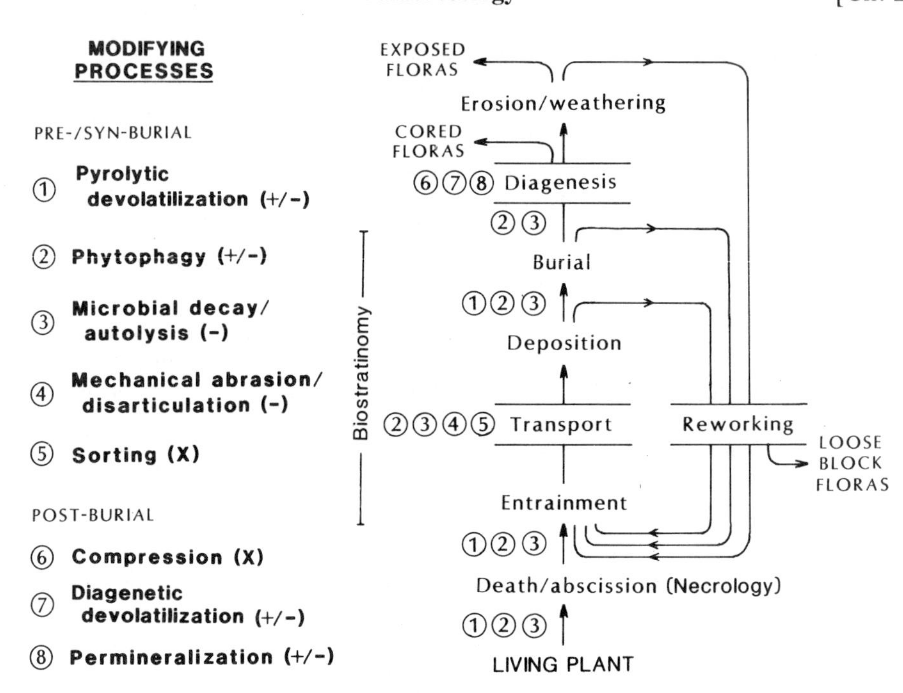

Fig. 2.4 — Conceptual taphonomic history of plant debris. Effect of modifying process on the overall probability of preservation: +, increased, −, decreased; ×, preservationally neutral.

devolatilizes plant tissue (Cope & Chaloner 1985, Scott 1989); the resulting pyrofusain (fossil charcoal) is simultaneously rendered inert, thereby increasing its resistance to chemical and microbial degradation, and brittle, thereby decreasing its resistance to mechanical abrasion. Much of our knowledge of taphonomic cycles and biases is derived from recent empirical studies of modern plant communities, focusing either on potential megafossils (e.g. Spicer 1981, 1989a, Scheihing & Pfefferkorn 1984, Ferguson 1985, Gastaldo 1985a, 1988, Spicer & Greer 1986, Spicer & Wolfe 1987, Burnham 1989, Burnham *et al.* in press, Behrensmeyer *et al.* in press) or potential microfossils (Farley 1988, Holmes in press). These are used as an interpretative framework for palaeobotanical assemblages. Although valuable, this uniformitarian approach has limitations when applied to communities as ancient as those of the Palaeozoic; the most effective studies use taphonomy as a vehicle to address specific palaeoecological questions.

2.2.2 Necrology and biostratinomy

Many plant organs, especially leaves and disseminules (dispersed isospores, megaspores, seeds, fruits, vegetative organs capable of independent existence), are routinely abscised during the growth cycle of the plant (it could be argued that Isaac Newton is the father of taphonomy, given his historic encounter with an abscised apple). Such physiologically mediated losses tend to be more periodic (typically seasonal) than pathogenic losses resulting from disease or herbivory. Even more traumatic losses caused by severe environmental perturbations (e.g. drought, severe storms, wildfire) are generally less frequent and less periodic, though monsoons are an exception to this rule. At best, such catastrophes liberate organs not usually considered ephemeral; at worst, they cause widespread mortality, releasing much biomass for potential fossilization.

Biostratinomy has provided taphonomy with its two most frequently used and palaeoecologically powerful terms, autochthony (untransported) and allochthony (transported). In a review of plant taphonomy, Gastaldo (1988: 18) offered a mainstream definition of autochthony as burial in the site of growth, thereby including both *in situ* proximal organs and aerial organs occurring as litter in the vicinity of the parent plant. He defined plant parts transported less than 500 m from the parent plant as hypoautochthonous and those transported more than 500 m from their source as allochthonous. Although plant parts can be readily assigned to these categories during studies of extant systems by tracing the passage of individual organs across landscapes and through a taphonomic system (e.g. Spicer 1981, Scheihing & Pfefferkorn 1984), the difficulties of applying these definitions to fossil plant assemblages extracted from restricted, typically two-dimensional, outcrops are obvious. As defined above, *in situ* organs are readily identified as autochthonous but there is no obvious means of distinguishing autochthonous, from hypoautochthonous litter. Moreover, the 500-m threshold separating autochthony from hypoautochthony is not only undemonstrable but also arbitrary. I suggest that a far more biologically meaningful threshold is whether disarticulated plant-parts remain within the community of origin or are transported beyond its boundaries (I will postpone discussion of criteria for locating such a threshold).

Thus, in this paper I redefine *autochthony* to encompass only organs that are demonstrably *in situ*, and discard hypoautochthony. I coin the term *parachthony* (Gr.: para=beside or nearby, cthonos=soil) for disarticulated plant-parts occurring within their community of origin, and redefine *allochthony* to encompass plant-parts transported beyond their community of origin (note that both parachthonous and allochthonous plant-parts are transported in the sense of having passed through a hydraulic medium). These distinctions emphasize the importance of delimiting communities and identifying their boundaries (ecotones). Although these pivotal terms tend to be used indiscriminately for entire fossil plant assemblages, strictly they describe individual plant-parts (consider, for example, a Palaeozoic palaeosol containing autochthonous, *in situ* stumps, capped by parachthonous litter and containing allochthonous, wind-dispersed saccate spores). Even wholly autochthonous fossil assemblages are generally enclosed in dominantly transported inorganic clasts; only pure authigenic limestones and some coals can be regarded as entirely untransported.

2.3 EVOLUTION OF LAND-PLANT CLADES AND COMMUNITIES THROUGH THE PALAEOZOIC

This account of land-plant evolutionary patterns and processes through the Palaeozoic is necessarily brief and simplistic. More comprehensive coverage can be found in excellent compilation volumes based on short-courses held in Texas in 1986 (Gastaldo & Broadhead 1986) and Virginia in 1987 (Evolution of Terrestrial Ecosystems

Consortium, in press), together with several valuable reviews (Banks 1968, Chaloner & Sheerin 1979, Edwards 1980, Scott 1980, 1984, Gensel & Andrews 1984, Edwards & Fanning 1985, Gensel 1986, Knoll *et al.* 1986, Collinson & Scott 1987a, b, Chaloner 1988, Taylor 1988, Crane 1989, Selden & Edwards 1989). The discussion is limited to the Eurameria Palaeokingdom (see section 4.2), since the various case-studies reviewed later in the chapter (sections 2.4–2.9) occur exclusively in that phytochorion.

Most of the following discussion focuses on the class- and order-level taxa depicted in Fig. 2.5. Although broadly based on Stewart (1983: 22), this classification is unconventional, most notably in the treatment of the angiosperms as a lowly order of the class Gymnospermopsida. Colloquial derivations of formal class and order names consistently bear suffixes of '-opsid' and '-alean' respectively, following Bateman (1990); this emphasizes the taxonomic rank assigned to the group and is highly relevant to arguments presented later in this paper. The stratigraphical ranges and evolutionary relationships of classes (and selected orders within the class Gymnospermopsida) depicted in Fig. 2.5 are highly tentative, as is the evolutionary integrity of most of the taxa, which has not been tested using cladistic methods (e.g. Wiley 1981). Of the classes shown, only the Lycopsida and Gymnospermopsida have strong claims to each being a clade: an evolutionary entity consisting of an ancestral species and all of its descendants. I shall return later to the relationship between phylogeny and palaeocommunity evolution.

2.3.1 Four phases of evolution in the plant kingdom

As a gross generalization, I would argue that the evolution of plant life can be broken down into four successive (albeit temporally overlapping) phases of evolutionary innovation, each defined by a suite of crucial advances in particular categories of biological phenomena.

The *biochemical* phase characterized the *c.* 2000 Ma history of life prior to the Ordovician. During this period, fundamental biochemical pathways such as those facilitating photosynthesis and respiration were established in anatomically simple cyanophytes and algae occupying aqueous environments. Subsequent biochemical innovations, such as the divergence of groups specializing in the C_3, C_4 and CAM modes of CO_2 assimilation (Spicer 1989b), were relatively trivial in nature (though not necessarily in ecological consequences). Also relevant was the evolution of sexual reproduction, entailing meiosis and zygote formation. These crucial events, well reviewed by J. W. Schopf (1970, 1975), Knoll (1985b, 1987), and Strother (1989) are beyond the scope of this chapter.

The *anatomical* phase spanned the Ordovician and Silurian. Here, erstwhile pioneering land plants struggled with the physical and physiological problems of adapting to a terrestrial existence, competing with the environment rather than with each other. At this time, most of the major tissue types evolved, together with the alternation between independent sporophyte and gametophyte generations that defines the pteridophytic life history.

The *morphological* phase reached an acme in the Devonian. At this time, land plants experimented with different arrangements of tissue types, apparently indulging in morphological and architectural escalation. This greatly increased the range

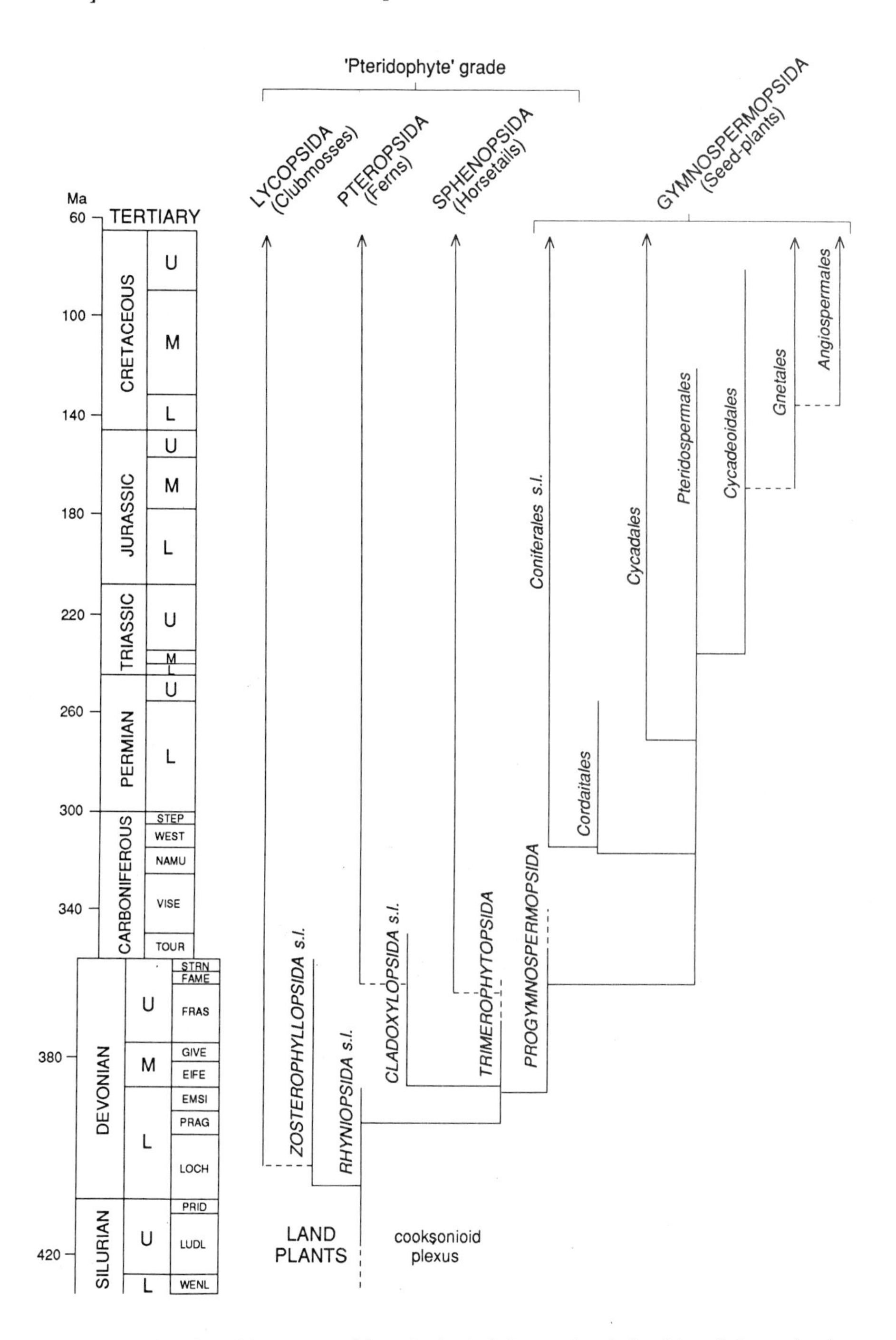

Fig. 2.5 — Stratigraphic ranges and hypothesized phylogenetic relationships of classes (capitals) and gymnospermous orders (lower case) of vascular plants. Note that the time-scale for the pre-Carboniferous is double that for subsequent periods. (Time-scale based on Leeder (1988) for the Carboniferous and Harland *et al.* (1989) for the remainder.)

and maximum complexity of gross morphological form, as well as the maximum body sizes of individuals found in different clades.

The *behavioural* phase increased exponentially through the Carboniferous and Permian, building on the previous phases and establishing the ecosystem dynamics that continue to control modern vegetation. Although interactions between individual plants and their abiotic environment remained important, they were increasingly supplemented with interactions among individuals. Differences in the phylogenetic relationships of the individuals involved delimit three successive subphases of interaction: plants of the same species, plants of different species, and plants with animals. Biotic competition for resources dominates the first two subphases, but the third is more diverse: it includes animal-mediated pollination and propagule dispersal, as well as herbivory. Morphological expression of behavioural innovation tends to be relatively restrained; a good (albeit probably post-Palaeozoic) example is the origination of the angiosperms, where major advances in pollination biology allowed unprecedented magnitudes of speciation, radiation and ecological dominance in most habitats but involved the production of no new tissue types and only minor morphological modification (the closure of the carpel surrounding the ovules).

Overall ecological complexity increased as each phase of innovation was superimposed on the products of previous phases. Anatomical and morphological innovations are more readily discerned in the plant fossil record than are biochemical and behavioural advances. Fortunately, it is the former that characterize the evolution of Palaeozoic vegetation and are thus most relevant to this chapter.

2.3.2 Ordovician and Silurian

Although ocean-marginal terrestrial environments may have supported microbial mats since the late Precambrian (e.g. Wright 1985), bryophytes were probably the first true land plants (embryophytes). Mishler & Churchill (1985) argued that they were derived from subaqueous charophytes and in turn gave rise to one or more clades of 'vascular plant' (here I use the term in its broad sense of pteridophytic plants with any form of conducting tissue). Some key adaptations to the physiological stresses associated with terrestrialization were biochemical, such as the use of the enzyme glycolate oxidase to overcome relatively high partial pressures of atmospheric oxygen and of flavonoids as UV screens (Chapman 1985). Other adaptations were physical, including desiccation-resistant materials enveloping spores (sporopollenin: Gray 1984, Raven 1985) and vegetative axes (cuticle). The advent of cuticle necessitated the evolution of stomatal pores for gaseous exchange; both cuticle and pores may have been co-opted, having originally evolved for functions different from those that they now fulfil (Knoll *et al.* 1986).

Although its origin is contentious, the early embryophyte life cycle, characterized by alternation between a haploid gametophyte generation and a diploid sporophyte generation, allowed retention of the egg (and fertilized zygote) within gametophytic nutritive tissues during its development. Genetic decoupling of the gametophyte and sporophyte generations also allowed heterozygosity and morphogenetic differentiation. The physiological and physical decoupling of the generations inherent in the derived pteridophytic life history allowed further differentiation, though some early land plants may have possessed morphologically identical (isomorphic) sporophyte and gametophyte generations (Remy & Hass 1991, Kenrick & Crane in press). Later differentiation favoured the increasingly dominant sporophyte generation, which is much more commonly fossilized than the gametophyte (Knoll *et al.* 1986). Pteridophytes still required the presence of free water for fertilization.

Increase in the body size and stature of sporophytes facilitated spore dispersal but also required specialized water-conducting tissues; these probably arose independently in several early lineages (Kenrick & Crane in press). The transition from passive to active water economy and increased need for nutrients also necessitated specialized absorbant structures penetrating the soil; initially these were rhizoids but later true roots appeared, perhaps rapidly acquiring symbiotic associations with mycorrhizal fungi (Beerbower 1985).

Evidence for Ordovician land plants is circumstantial, notably the putative palaeosols (Feakes & Retallack 1988) and enigmatic spores formed in unusual configurations (e.g. Gray 1984, 1985). Other plant remains of this age, notably putative cuticular fragments and isolated conducting cells, are fragmentary and present problems of identification (of both function and taxonomic affinities); moreover, they are typically preserved sparingly and in marine facies, hindering palaeoecological interpretation (Knoll *et al.* 1986). All Ordovician remains may be bryophytic *s.l.* By the Wenlock, meiotic triradiate spores had appeared and begun to diversify (Pratt *et al.* 1978, Gray 1985, Richardson 1985); together with fragmentary megafossils (Edwards & Fanning 1985), these suggest that pteridophytic plants had evolved (Fig. 2.5). The Ludlow and Pridoli are marked by the radiation of the 'cooksonioids,' a persistently enigmatic but evolutionarily crucial plexus of land plants (D. Edwards 1980, Edwards *et al.* 1983, Edwards & Fanning 1985, Chaloner 1988, Taylor 1988). They apparently possessed well-differentiated tissues but were small-bodied and architecturally simple, being composed of naked determinate axes. Many were rhizomatous, forming extensive monospecific clonal 'turfs' that obtained nutrients via shallowly penetrating rhizoids (Tiffney & Niklas 1985). Spore production generally appears sporadic and probably served primarily to enable colonization of new areas.

Although the binding properties of the rhizoids were weak, soils became more complex and better developed, supporting substantial arthropod faunas (Retallack 1986); other arthropods occupied the litter layer, where they specialized in detritivory or predation (Rolfe 1985), implying the emergence of simple food webs. Most plant fossil assemblages of this age are from the palaeoequatorial belt and allochthonous, occurring in fluvial and ocean-marginal facies. This suggests association with surface water bodies in tropical and subtropical climates (Andrews *et al.* 1977, Edwards *et al.* 1983, DiMichele *et al.* in press).

2.3.3 Lower Devonian

Many more plant fossil assemblages are known from the Lower Devonian, though again most are near-equatorial, suggesting a warm and at least seasonally arid climate. Also, they continued to be dominantly allochthonous and associated with ocean-marginal (often marine) deposits (Gensel 1986). The Gedinnian marked the origination from the cooksonioid plexus of three major classes (Fig. 2.5): the rhyniopsids *s.s.* (Kenrick *et al.* 1991, Kenrick & Crane in press), the zosterophyllopsids, and the lycopsids (Niklas & Banks 1990, Hueber in press). These were followed in the Siegenian and Emsian by the trimerophytopsids; the emergence of these four classes constitutes a major phylogenetic and architectural radiation (Chaloner & Sheerin 1979, Banks 1980, D. Edwards 1980, Gensel & Andrews 1984, Knoll *et al.* 1984, Knoll 1985a, Gensel 1986).

Growth remained rhizomatous in most species, but true roots evolved in several higher taxa, allowing access to groundwater and providing firmer anchorage. This in turn enabled root-bearing plants to modify their environment of growth, promoting pedogenesis and stabilizing the resulting soils against subaerial and subaqueous erosion (Remy *et al.* 1980, Retallack 1985). Above ground, determinate lateral emergences (enations) appeared in the zosterophyllopsids, lycopsids and trimerophytopsids. The contributions of enations to photosynthesis and gaseous exchange would have been minimal, implying that they were mechanical rather than physiological adaptations. Together, these innovations would have promoted expression of species-specific microhabitat preferences (Knoll 1985a, Gensel 1986).

2.3.4 Middle and Upper Devonian

Data are regrettably poor for the Middle Devonian in general and Emsian Stage in particular (Gensel 1986). By the Eifelian, the cooksonioid complex, rhyniopsids and zosterophyllopsids had yielded to advanced trimerophytopsids and new classes/orders derived either directly or indirectly from the trimerophytopsids; these included the cladoxylopsids–iridopteridaleans, hyenialeans, and progymnospermopsids (groups often implicated in the origins of the pteropsids, sphenopsids, and gymnospermopsids respectively). They inherited the pseudomonopodial growth, strongly unequal branching and megaphylls (leaves supposedly derived from branching systems: e.g. Stewart 1983) of the advanced trimerophytopsids, which allowed increased stature and architectural specialization. By the Givetian, secondary thickening (wood production) had evolved, apparently independently in at least three classes: the lycopsids, cladoxylopsids and progymnospermopsids (Fig. 2.5). These innovations yielded taller, larger-bodied species able to exploit the third (vertical) spatial dimension; thus, clonal turfs were supplemented with scrubby 'protoforests', some composed of the determinate, rhizomorphic tree-lycopsids that would eventually dominate many late Devonian and Carboniferous swamp communities. Vegetational stratification in some Frasnian communities reached a scale comparable with that of extant forests (Fig. 2.6: see also Niklas 1982, Scheckler 1986a).

Heterospory allowed concentration of food reserves in the 'female' spore and thereby improved its probabilty of successful establishment following fertilization, albeit often at the cost of average dispersal potential. This phenomenon is perceived as a key precursor to the ecologically successful seed habit (e.g. DiMichele *et al.* 1989). Although putative heterosporous progymnospermopsids occur as early as the Emsian (Andrews *et al.* 1974), their radiation is documented in Givetian and Frasnian strata. Both aneurophytalean and archaeopteridalean progymnospermopsids diversified and species became more ecologically specialized, the latter increasing in dominance at the expense of the former through the Frasnian, spanning a

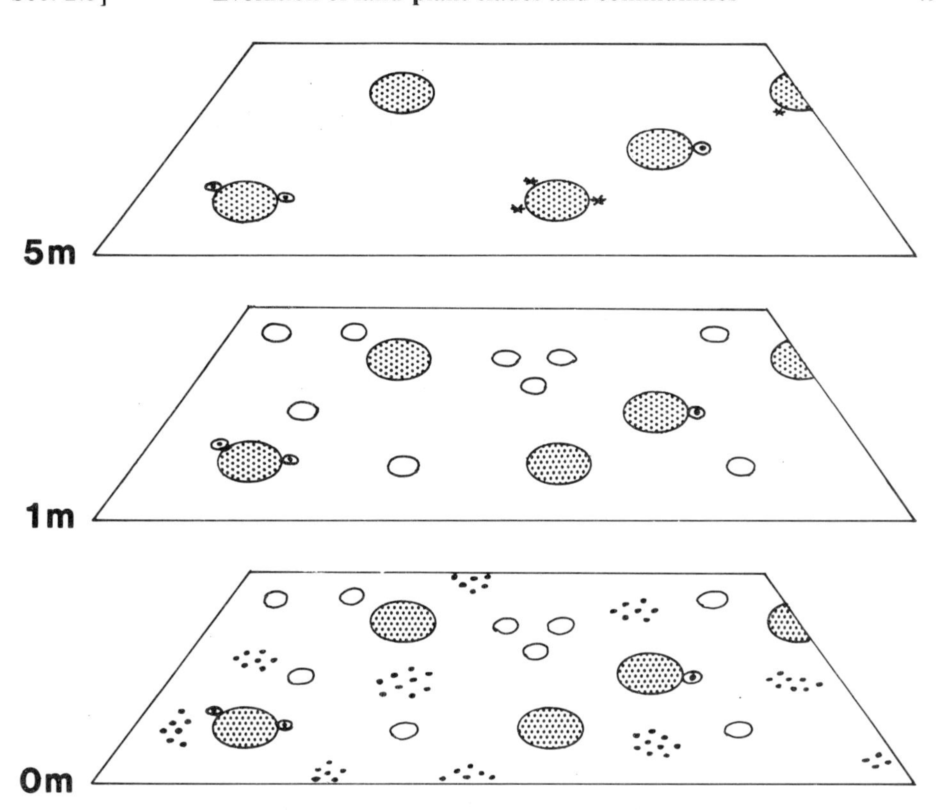

Fig. 2.6 — Conceptual representation of stratification in a forest community. Small dots=terrestrial herbs, asterisks=epiphytes, bulls-eyes=lianas, circles=shrubs, stippled circles=trees.

recognized period of extinctions (Scheckler 1986c). By the mid-Frasnian, some archaeopteridalean species formed monocultures on levees and floodplains (Retallack 1985). They were large trees bearing substantial, complex and disposable branches (Retallack 1985, Scheckler 1986a). At the same time, large-bodied lycopsids and putative sphenopsid precursors became important components of wetland communities (Matten 1974, Gensel 1986). Increases in vegetational stratification and physiological efficiency facilitated niche partitioning, but also greatly increased litter density on the soil surface (and thereby increased the risk of wildfire).

By the Famennian, heterospory had evolved independently in three lineages; it was merely an imperfect experiment in the barinophytalean derivatives of the zosterophyllopsids, but became increasingly canalized and sophisticated in the lycopsids and progymnospermopsid–gymnospermopsid clade. Archaeopteridalean progymnospermopsids and the enigmatic 'pre-fern' *Rhacophyton* remain dominant in most Famennian and 'Strunian' assemblages (Scheckler 1986b, c). Wetland assemblages of this age are composed of tree-lycopsids, sphenopsids and the earliest known seed plants, together forming perhaps the first communities well-adapted to soils deficient in oxygen and nutrients (Scheckler 1986a, c). The pteridospermalean seed plants are regarded as having surmounted a major ecological threshold; the seed habit liberated them from dependence on free water for fertilization, permitting the invasion of drier habitats (though it is interesting to note that the early seed plant assemblages occur in wetlands: Fairon-Demaret 1986, Rothwell & Scheckler 1988, DiMichele *et al.* 1989). Pteridospermalean diversity increased during the Strunian, while end-Strunian extinctions eliminated the archaeopteridaleans (Matten *et al.* 1984, Scheckler 1986a, c).

To summarize, by the end of the Devonian, several classes of plant that evolved during the Late Silurian or Early Devonian had become ecologically insignificant or extinct, and ecological dominance had passed to the four classes that have left extant descendants: the lycopsids (clubmosses), sphenopsids (horsetails), pteropsids (ferns), and gymnospermopsids (seed plants) (Fig. 2.5). Ecological preferences traditionally attributed to these classes in the Late Carboniferous were already evident; wetlands supported tree-lycopsids, floodplains yielded pteridospermaleans, pteropsids and smaller-bodied lycopsids, and riparian habitats were often sphenopsid monocultures. Hence, plant communities were structurally modern in concept if not in detail. Although arthropods diversified rapidly through the Middle and Upper Devonian (Kevan *et al.* 1975, Rolfe 1980, Shear *et al.* 1984, 1989), they remained carnivores or detritivores, eschewing the more intimate interactions with plants necessary to drive coevolution (DiMichele *et al.* in press).

2.3.5 Lower Carboniferous

Recent discoveries have pushed the origins of several important taxa back from the Early Carboniferous to the Late Devonian. Nonetheless, the Early Carboniferous remains an important period of taxonomic and ecological diversification. The Late Devonian and Tournaisian are implicated as the period of most rapid architectural and ecological radiation of the tree-lycopsids (Bateman *et al.* in press) and of the calamopitid and lyginopterid pteridospermaleans (Long 1975, Pfefferkorn & Thomson 1982, Rothwell & Scheckler 1988, DiMichele *et al.* in press). A late Tournaisian assemblage of seed-like lycopsid propagules (Long, 1968) suggests that the pinnacle of tree-lycopsid specialization, *Lepidophloios*, had already evolved. Heterospory appeared in at least one early species of sphenopsid (Bateman in press), leaving the pteropsids as the only uniformly homosporous tracheophyte class still in existence. The high canopy niche vacated by the archaeopteridaleans was filled by large-bodied pteridospermaleans (DiMichele *et al.* in press). Diversification of several pteropsid orders and families occurred throughout the Tournaisian and Visean (Galtier & Scott 1985).

By the end of the Visean, most major growth habits (tree, shrub, herb, liana) were represented and forest stratification was well-established (Scheckler 1986a, DiMichele *et al.* in press). This paved the way for more complex seral successions (Bateman & Scott 1990). Rapid species turnover in pteridospermaleans, a feature of the Lower Carboniferous (Scott *et al.* 1984), continued into the Namurian. Limited studies suggest that Early Carboniferous arthropods resembled those of the Late Devonian. Tetrapods were represented mostly by amphibians, who fed on each other or on arthropods; thus, the evolution of terrestrial plants remained largely independent of the evolution of terrestrial animals (DiMichele *et al.* in press).

2.3.6 Upper Carboniferous

The Late Carboniferous saw the zenith of peat formation in the extensive pan-equatorial coal-swamps of Laurasia, which formed largely on coastal plains and deltas (Phillips & Peppers 1984, DiMichele *et al.* 1986, Mapes & Gastaldo 1986) during a period when the equatorial rain belt was unusually broad (Ziegler 1990). Coal-bearing strata have yielded the most complete record of any Palaeozoic vegetation type, though it is also a record highly biased towards hydrophilic communities.

In many such communities from the Westphalian, lycopsids were the largest trees and generated the greatest biomass (Phillips *et al.* 1977, DiMichele & Phillips 1985, Cleal 1987, Phillips & DiMichele in press). Common subcanopy trees in these bizarre forests were large-leaved medullosan pteridospermaleans, while at least some of the lyginopterid pteridospermaleans grew as vines and lianas. Pteropsids spanned a wide range of growth habits: small trees, scramblers, vines, herbs, and putative epiphytes (Trivett & Rothwell 1988, Lesnikowska 1989, Rothwell in press; Fig. 2.6). Sphenopsids were represented by dense clonal thickets of sphenophyllaleans and typically riparian stands of tall woody calamites (Hirmer 1927); both were well-adapted to survive flooding and limited burial. Cordaitalean seed plants reached an acme late in the Mid-Westphalian, encompassing a range of architectures and habitats that included understorey scrub and coastal mangrove analogues (Rothwell & Warner 1984, Mapes 1987).

Severe climatic drying at the beginning of the Stephanian eliminated most of the tree-lycopsids but allowed further increases in the maximum body sizes of calamites and marattialean tree-ferns. The few allochthonous assemblages attributed to extrabasinal habitats reveal an increasingly broad range of relatively xeromorphic species very different from those of the wetlands. Shrubby coniferaleans and cordaitaleans were increasingly important components (Scott & Chaloner 1983, Mapes & Gastaldo 1986, Mapes 1987, McComas 1988, Rothwell & Mapes 1988), together with enigmatic pteridospermaleans and putative cycadopsids (Read & Mamay 1964, Leary & Pfefferkorn 1977, Leary 1981, Galtier & Phillips 1985, Kerp *et al.* 1989). At least one coniferalean shows the earliest convincing evidence of true seed dormancy (Mapes *et al.* 1989), rather than the delayed fertilization following pollination that may have mimicked dormancy in some pteridospermaleans (DiMichele *et al.* in press). Dormancy would have been especially valuable in water-stressed, fire-prone habitats.

The coal-swamp ecosytems supported rich arthropod faunas (Scott 1980, Scott & Taylor 1983, Rolfe 1985, Labandeira & Beall 1990). Use by a few plant species of arthropods as pollen and/or propagule vectors has also been suggested, though this is difficult to demonstrate with confidence. Arthropods that consumed plant material increased in mean body size, and insects in particular developed a range of niches comparable with those of present-day faunas, though assemblages continued to be dominated by detritivores. Coprolite evidence indicates the presence of specialized wood-borers, while other arthropods consumed the increasingly widespread bio-degradative fungi (Stubblefield *et al.* 1983). Spores and seeds were probably consumed preferentially, though the appearance in plants of various types of putative protective tissues, and some evidence of leaf chewing, leaf mining and piercing-and-sucking together suggest limited herbivory (Labandeira & Beall 1990). However, the Late Carboniferous diversification of amphibians and early reptiles yielded specialist herbivores only in the more xeric habitats of the Stephanian (DiMichele *et al.* in press).

2.3.7 Permian

During the Early Permian, climate became increasingly continental (Parrish *et al.* 1988) and habitats increasingly heterogeneous (Havlena 1970). Dry habitats that had been marginal in Late Carboniferous Laurasia became predominant (Mapes & Gastaldo 1986). The xeromorphic and mesomorphic seed plants that appeared in the Late Carboniferous gradually extended their ecological and thus geographical ranges, while the classic swamp-dwelling 'pteridophytes' declined precipitously. Seed-plant radiations generated new groups of cycadopsids, coniferaleans and enigmatic pteridospermaleans such as the peltasperms and gigantopterids; all apparently showed relatively high degrees of endemism (Read & Mamay 1964, Mamay 1967, 1976, Meyen 1982, Kerp & Fichter 1985, Mapes & Gastaldo 1986). Floras reminiscent of the Stephanian wetlands persisted only rarely, in reliably wet soils (although they persisted through much of the Permian in Cathaysia — see section 4.5.2).

As in the Upper Carboniferous, Permian higher taxa can be arranged in a cline of soil moisture preference, ranging from xerophytic coniferaleans to hydrophytic sphenopsids and lycopsids (DiMichele *et al.* in press). Thus, the rise to global dominance of seedplants, which prompted Kryshtofovich's (1957) erection of the Palaeophytic–Mesophytic transition in the Mid-Permian, was both ecologically and geographically diachronous (Knoll 1984, Meyen 1987, DiMichele & Aronson in press). It was accompanied by a major expansion in plant–animal interactions. Herbivory increased greatly among arthropods and tetrapods, conferring on food-webs a more modern aspect (DiMichele *et al.* in press). Biotically mediated pollination and propagule dispersal also probably increased in frequency; in short, the integrated 'behavioural' phase of terrestrial ecosystem evolution began in earnest.

From the Permian onward, only two taxonomic classes maintained substantial representation in land-plant communities: gymnospermopsids (seed plants) were generally dominant, with homosporous pteropsids (ferns) important in some habitats, particularly those subject to considerable stress but reliably providing (at least periodically) the free water necessary for pteridophytic reproduction. Thus, the Palaeozoic had finally set the pattern for the remaining 250 Ma of land-plant evolution and community dynamics (Fig. 2.5).

2.3.8 Summary of evolutionary trends and causal hypotheses

During the early Palaeozoic, the terrestrial realm was essentially an ecological vacuum, at most occupied only in ocean-marginal areas and only by bacterial and algal mats. By the end of the Palaeozoic, several groups of plants and animals had successfully colonized the land and become organized into well-integrated ecosystems that were structurally and dynamically comparable with those of today. A series of class- and order-level biotic replacements resulted in worldwide dominance of communities by the two classes of vascular plant that remain dominant in modern floras: the seed plants and, to a lesser degree, the ferns.

Several directional evolutionary trends are evident in that part of the fossil record that spans the 150–200-Ma period separating the origination of land plants from that of structurally modern ecosystems.

On average, land plants became better adapted to a terrestrial existence. This involved increases in anatomical, morphological and behavioural sophistication that generally (though by no means always) involved increased overall complexity. Land plants became more adept at the 'economic' (*sensu* Eldredge 1989) aspects of their existence, notably obtaining water and nutrients from the soil and increasing the efficiency of photosynthesis and respiration. Biomechanical innovations, such as the evolution of a hierarchy of indeterminate and determinate meristems (localized tissues that determine plant form during development) and the ability to produce secondary tissues, allowed much larger plant body-size and exploitation of the third spatial dimension, height. This caused stratification (Fig. 2.6), opening up communities for competition, a process that previously had been focused along the margins of clonal mats (Tiffney & Niklas 1985, DiMichele *et al.* in press).

Profound reproductive innovations included the evolution of life histories involving alternation of gametophyte and sporophyte generations, the ensuing economic dominance of the sporophyte generation, partitioning of nutrients among spores in favour of the 'female' megaspore (heterospory), and retention of a single functional megaspore (now largely enclosed by protective sterile tissues) on the sporophyte until pollination had been effected. Current evidence suggests that all of these key functional innovations may have been iterative; for example, heterospory and secondary thickening evolved independently in all four of the classes that persisted beyond the mid-Carboniferous boundary (Fig. 2.5). Opportunistic clonal species declined in importance, yielding to species that were on average more *K*-selected (Pianka 1970) and dependent on synchronized sexual reproduction.

Economic and reproductive innovations together allowed colonization of ever-increasing ranges of habitats. Plants showed increased tolerance to low soil moisture, extremes of temperature, and severe habitat disturbance. Consequently, the range of ecological vacuums still available for unopposed invasion gradually diminished. Together with increasingly constrained (canalized) plant development, this progressively decreased the probability of establishing radically novel, mutationally generated phenotypes, which are inevitably poorly adapted until they establish populations capable of being honed to increased fitness by natural selection (Valentine 1980, Arthur 1984, Knoll 1986, DiMichele *et al.* 1989, Bateman *et al.* in press). Thus, increased ecosystem complexity and connectance may have constrained further radical evolutionary experimentation in an inevitable negative feedback loop.

Early land plants competed primarily with the environment; any biotic interactions tended to be confined to conspecific individuals comprising a local population. Consequently, the potential (fundamental) niche of a species, delimited by its environmental tolerances, approximated its actual (realized) niche (Hutchinson 1965). However, as biodiversity increased, the range of vacant habitats declined. Once a habitat was occupied it could not readily accommodate newly evolved species (see Harper 1977). Accommodation could potentially be achieved through biotic competition, either by displacing the occupant (thus causing its migration or local extinction: extirpation) or by partitioning resources so that the realized niches of both competing species are reduced in scope relative to their fundamental niches; they can then co-exist. Alternatively, invasion can be opportunistic, taking advantage of empty niches temporarily vacant due to environmental perturbations such as fire, flood or climate change. Under these circumstances, the initial recolonist may not be the species best adapted to live in the habitat, but merely the first to reach the site. The relative importance of biotic and abiotic stresses in promoting vegetational change have been much debated (e.g. Valentine 1980, Vermeij 1987, DiMichele *et al.* 1987, in press).

Long-term, large-scale vegetational changes probably largely reflect regional or global climate change; this is especially likely for the post-Westphalian transition associated with the increased aridity and seasonality that resulted from cessation of the Gondwanan glaciations and assembly of the Pangaean supercontinent (Raymond *et al.* 1985, Ziegler 1990). In contrast, biotic interactions underpin the influential theory of escalation — adaptive evolution necessary to keep pace with increased competition and environmental degradation (Van Valen 1973, Valentine 1980, Vermeij 1987). Escalation and increased diversity together may have reduced the differential fitness among potential competitors for a particular niche, thereby reducing the risk of displacive invasion (Knoll & Niklas 1987).

Biotic interactions undoubtedly increased through the Palaeozoic, not only plant–plant competition but also plant–animal interactions, both parasitic (e.g. herbivory) and mutualistic (e.g. faunal pollen vectors). Such co-evolutionary relationships were slow to develop; those with arthropods and later with tetrapods were especially important contributors to plant evolution (Beerbower 1985, DiMichele *et al.* in press). This increased the ecological linkage (connectance *sensu* Pimm 1984) among the species comprising a community, and thereby placed more emphasis on synecological relative to autecological factors.

In terms of modern ecological theory, early land-plant communities were undoubtedly strongly Gleasonian; local vegetation type reflected individualistic responses of species to extrinsic (dominantly abiotic) factors, typically of large (e.g. global/regional climate) or intermediate (e.g. wildfire, flooding, mass-flow) scales. Such communities are ad hoc associations of species sharing broadly similar environmental requirements; they readily exhibit a plastic response to extrinsic stresses, the effect being proportional to the severity of the environmental change. However, through the Palaeozoic, communities became increasingly Clementsian. Intraspecific interactions increased in frequency and importance, binding communities together and increasing their likelihood of long-term persistence through time; some Late Palaeozoic communities apparently persisted, essentially unchanged, for several million years (DiMichele *et al.* in press). Thus, they became potential evolutionary units. Such communities resist extrinsic stress until they exceed the threshold level at which behavioural linkages are broken; a brittle response may then ensue as the ecosystem suddenly collapses (DiMichele *et al.* 1987, DiMichele & Aronson in press). The degree and importance of biotic interactions at any particular time in the Palaeozoic remains highly contentious. It is clear that the evolution of the interactions (particularly those between plants and animals) lagged well behind that of the interactors (DiMichele *et al.* in press), even though vascular plants had long played a key role as primary producers in terrestrial food pyramids.

The above scenario of community evolution is both logical and predictable from current evolutionary theory — but how closely does it approximate the truth? In the following section, the above synthesis is contrasted with case studies summarizing some of the best available data; do they fit the scenario comfortably, or are they being forced unjustifiably into preconceived hypotheses?

II. Grim realities

In section I, I have outlined the nature of the plant fossil record and distilled current opinions on the major events in early land-plant evolution. I will now present a series of case studies summarizing several of the most rigorous palaeoecological investigations of Palaeozoic communities thus far published (all are from the Eurameria Palaeokingdom). I shall focus on the relationship between the methods used and the interpretations placed on the results, noting what I perceive as the pros and cons of each study. The concluding synthesis highlights discrepancies between available data and assertions, and suggests how the gap between knowledge acquired and knowledge desired could be narrowed in future studies.

2.4 TARGROVE QUARRY, ENGLAND (LOWER DEVONIAN, ADPRESSION)

The early history of terrestrial plant life is documented largely in adpression assemblages. Although generally infrequent and widely dispersed, some are concentrated in areas highly conducive to palaeoecological interpretation (e.g. the Gaspé coast of Quebec). The paucity of rigorous integrated studies of such localities is vexing, leaving this most crucial period of land-plant evolution undesirably prone to poorly substantiated speculations.

The following discussion is based largely on Edwards & Fanning's (1985) detailed palaeobotanical study of Targrove Quarry, near Ludlow, western Central England. The sediments are Gedinnian (*c.* 400 Ma following Harland *et al.* 1989) and were deposited as the distal alluvial facies of the Lower Old Red Sandstone, close to the south-east margin of the Laurasian continent at a palaeolatitude of 20–25°S. The sequence is rich in plant remains, dominantly devolatilized adpressions. Although the fossils are deficient in anatomical detail, acetate pulls have yielded some poorly preserved cuticle, spores and putative tracheids.

Probable non-vascular plants include the large putative fungus *Prototaxites* (Hueber in prep.), the abundant but enigmatic cuticularized *Nematothallus* (D. Edwards 1982), and another putative thallophyte, *Pachytheca*. The abundant axial fragments of cooksonioids and other, demonstrably vascular species rarely exceed 2 mm in diameter and 30 mm in length. A few specimens show isotomous branching (*Hostinella* sp.), K-branching (cf. *Zosterophyllum fertile*), and lateral serrations (Edwards & Fanning 1985, plate 1), but most are unbranched and unornamented. Consequently, classification and identification of the plants emphasize the gross morphology of axial fragments (most short and unbranched) that bear sporangia. Species are distinguished by sporangium characters (size, length : width ratio, presence of a distal border, shape of the proximal margin), together with the degree of attenuation of the subtending axis and the ornamentation of any *in situ* spores. Equality and frequency of branching of the subtending axes are also important, but rarely shown by the fragmentary Targrove material (Lang 1937, Edwards & Fanning 1985).

Six form-species bear unbranched sporangia that are consistently broader than long. Two bear small (less than 2.4 mm) sporangia that resemble *Cooksonia caledonica* but terminate lateral branches. They are sufficiently complete to demonstrate overtopping, and are thus assigned to *Renalia* spp. Four other Targrove species with sporangia 1.2–2.4 mm in largest diameter remain assigned to *Cooksonia: C. hemisphaerica, C. pertonii, C. cambrensis*, and cf. *C. caledonica*. Two spore 'genera' occur independently in different specimens of *C. cambrensis* (Fanning *et al.* 1988), suggesting that it is a form-species. Fusiform unbranched sporangia considerably longer than wide range in size from 4.0×0.7 mm to 0.7×0.3 mm. Most are attributed to the form-genus *Salopella*; different *in situ* spores suggest the presence of at least two species. A few specimens with clusters of proximally fused elongate sporangia are described as cf. *Yarravia* sp. Ellipsoidal unbranched sporangia range from 2.6×1.9 mm to 0.7×0.4 mm and vary in apex shape, sporangium wall anatomy, and type of *in situ* spores. They represent at least two species of a new genus. Finally, a new monotypic genus possesses blunt-tipped, bilobate sporangia that contain tetrahedral spore tetrads.

As a result of indifferent preservation, Edwards & Fanning (1985) could assign only half of the several hundred sporangia recovered to any of the above form-species. The presence of about 12 coarsely defined form-species renders Targrove the most diverse assemblage of cooksonioids known. In contrast, the early zosterophyllopsids are represented only by two equivocal specimens. Interestingly, associated assemblages of dispersed miospores are much more diverse than can be accounted for by spores found *in situ* in the cooksonioids; bryophytes may have contributed some of the excess form-species.

Repeated failure to reproduce Lang's (1937) reported extraction of tracheids from *Cooksonia hemisphaerica* (in fact extracted from associated sterile axes of uncertain identity), and indeed from any of the Targrove Quarry cooksonioids, casts doubt on their status as vascular plants. D. Edwards (1980) and Edwards & Fanning (1985) tentatively supported Niklas's (1976) hypothesis that lignin biosynthesis evolved early in the history of plant life, preceding the evolution of the tracheids that delimit bona fide vascular plants (e.g. Kenrick & Crane 1990, in press). Initially, lignin may have been deposited in peripheral tissues, providing structural support. Lignified conducting tissues were a later innovation, and may have evolved iteratively in several cooksonioid lineages (e.g. Taylor 1988).

Detailed lithological logs of the Targrove section have not been published. Edwards & Fanning (1985) described the sequence as a conglomerate that is overlain by channel-fill sandstones and passes laterally into floodplain deposits. Severe disarticulation and sorting suggest that all of the plant remains are highly transported and presumably allochthonous. *Prototaxites* clasts occur in the coarse basal horizons, abundant axial and sporangial fragments in the lower part of the channel-fill, and comminuted fragments associated with ripples in the upper part. Cuticular fragments dominate the floodplain debris. Larger and/or more proximal organs, including all rooting structures, are absent from Targrove; presumably they were deposited closer to the source communities. Thus, we lack knowledge of the overall morphology of the cooksonioids, and many palaeobiological assertions (e.g. of clonal growth: Tiffney & Niklas 1985) are speculative. Although probably exaggerated by taphonomic concentration, the abundance of reproductive material at Targrove supports Niklas *et al.*'s (1980) assertion that many cooksonioids were *r*-selected opportunists (i.e. quick growing and rapidly maturing, producing numerous propagules).

Impure limestone clasts in the section have been interpreted as reworked pedogenic calcretes. Uniformitarian extrapolation from modern calcretes, together with regional depositional patterns, indicates a warm to hot, highly seasonal climate. This implies either that the Targrove flora was xeromorphic or that it was associated with the larger, more persistent water-bodies on the fossil landscape (D. Edwards 1980, Edwards & Fanning 1985). The combination of apparently rapid deposition and considerable concentration of organic matter implies that vegetation close to the channel was dense, while the high species diversity relative to other early Devonian assemblages suggests that the channel sampled several 'microcommunities' within its catchment.

2.5 RHYNIE, SCOTLAND (LOWER DEVONIAN, PETRIFACTION)

The best-known terrestrial Devonian lagerstätten remains the classic, siliceously petrified Rhynie Chert biota from Scotland (e.g. Chaloner & Lawson 1985). These Old Red Sandstone strata were deposited along the eastern margin of the Laurentian continent at a palaeolatitude of 0–25°S. However, lack of exposure has seriously handicapped attempts to securely date the succession — it is currently considered middle Siegenian (*c.* 393 Ma; Westoll 1977) — and to interpret the taphonomic history of the fossils. After featuring in the classic monographs of Kidston & Lang (1917, 1920a, b, 1921a, b), the fossil plants attracted little empirical attention until radical taxonomic revisions of early land plants (e.g. Banks 1975) and increased interest in synecology and co-evolution (e.g. Kevan *et al.* 1975, Labandeira & Beall 1990) prompted re-examination of elements of the plants and associated fossil arthropods.

Although often discussed as though they were derived from a single flora, the Rhynie fossils occur throughout a 2.4-m-thick succession that consists of several organic-rich chert horizons (each 3–33 cm thick) intercalated with clastic sands of varying degrees of purity. The stratigraphical distribution of petrified species was described qualitatively in the text of Kidston & Lang (1921b: see Tasch 1957, Appendix 3) using the measured section of Tait (in Horne *et al.* 1916). These rarely cited but valuable data are summarized in Table 2.1. A more recent log based on a borehole is summarized by Trewin (1989) but full details have yet to be published.

Each assemblage is dominated by a single species of plant fossil; only in one horizon (6d′) do three of the four species present co-occur. *Rhynia* is confined to the lowermost horizon (1a), where it formed an *in situ* sward before being inundated with waters that allowed 'blooms' of algae and the small aquatic crustacean

Lepidocaris rhyniensis (Tasch 1957, Edwards & Lyon 1983). Stomata at the bases of the upright axes of *Rhynia* militate against prolonged inundation (Kidston & Lang 1921b, D. S. Edwards 1980). Dominance in the overlying horizons (1a′–6d′) alternates between *Asteroxylon* and *Horneophyton*; both apparently grew *in situ* in horizon 1c (Kidston & Lang 1921b). *Aglaophyton* occurs as occasional, apparently parachthonous, influxes in horizons 1–6, but following deposition of sandstone (7) it provides an uninterrupted series of monotypic assemblages (8–17), some possibly autochthonous. *Horneophyton* re-appears in the more clastic horizons (18–20) that terminate the plant-bearing sequence.

Tasch (1957) argued that the four dominant plant species differed primarily in tolerance to substrate moisture; *Rhynia* preferred the driest conditions and *Horneophyton* the wettest, with *Asteroxylon* and *Aglaophyton* favouring intermediate conditions. Clastic influxes, probably resulting from periodic flooding, disrupted the ecosystem and allowed re-colonization by whichever species was best adapted to the subsequent soil moisture regime. First *Rhynia* and then *Asteroxylon* failed to recolonize the area after such disturbance events, whereas *Horneophyton* returned from the preservational wilderness following a period of *Aglaophyton* monoculture (Table 2.1). The low diversity of each assemblage supports the growth model of

Table 2.1 — Stratigraphic distributions of dominant anatomically preserved plant species through the Rhynie Chert

Horizon(s)	Texture	Thickness (cm)	*Rhynia*	*Asteroxylon*	*Aglaophyton*	*Horneophyton*	Other groups
(21-22)	***	13					
20	**					*	
19	**	15			?*	+	
18a	*					*	F
18b	*	23		?+	+		
17	*					*	
16	**	30			*		
15	*	8		*			
14	*	8		*			
13	*	3		*			
12	*	8		*			
11a	*			?*			
11b	*	13		*		A	
10	*			?*			
9	**	33		*			
8	**			*			
(7)	***	8					
6d′	**		+	+	+		
6d	**		+		+		
6c	*				*		
6b	**	+		+			
6a	*	23		+	*		
(3–5)	***	c.20					
2	**	*					
1c	*		!*		!+		
1b	*			+	*		
1a′	*					*	
1a	*	c.33	!*		+		A F Ar

Parenthetic horizons lack identifiable plant remains. Textural classes: *, dominantly organic; **, intermediate; ***, dominantly clastic. Thicknesses include all horizons prior to the next value above.
Plant species frequency: +, present; *, dominant; !, probably *in situ*, ?, possibly *in situ*.
Other organisms: A, chlorophyte algae; F, fungi; Ar, arthropods. Data from Tait in Horne *et al.* (1916), Kidston & Lang (1917, 1921b), and Tasch (1957).

Tiffney & Niklas (1985); these rhizomatous species reproduced primarily by vegetative division. The resulting extensive clonal 'turfs' dissuaded colonization by other species unless disrupted by pronounced environmental perturbations.

Knoll (1985a) argued that the siliceous petrifaction at Rhynie is consistent with Leo & Barghoorn's (1976) model: partial degradation of the plant tissues lowers pH and frees hydroxyl and other functional groups for multiple hydrogen bonding with mono- or polysilicic acid in groundwater solution. Initial siliceous nucleation around the plants is followed by further precipitation of polymerized silicic acid. Silicification was incomplete at Rhynie, where amineralized voids have persisted in the matrix. Partial degradation of the plants is evidenced by decayed tissues and the occurrence of fungal hyphae within the plant axes (Kidston & Lang 1921b, Knoll 1985a). The Rhynie sequence has been interpreted as a series of wetland peats infiltrated by silica-charged waters from nearby volcanic springs (Kidston & Lang 1917, 1921b, D. Edwards 1980, Scott 1984), though evidence for the presence of springs, and for the extent and nature of the mire(s), requires critical re-evaluation.

Palaeobiological interpretations of the Rhynie assemblages are better-supported. Reconstructions of the plants by Kidston & Lang (1917 *et seq.*) have in turn been subject to rigorous revisions, notably by D. S. Edwards (1980, 1986, Edwards & Edwards 1986). Although hampered by the difficulty of obtaining satisfactory acetate peels and consequent reliance on petrographic thin sections, these studies have been aided by the often excellent anatomical preservation (e.g. Chaloner & MacDonald 1980), simple architectures and typically relatively mild disarticulation of the plant species, and by the low diversity of individual horizons (Table 2.1). The reconstructions have made Rhynie the single most influential locality in discussions of the evolutionary relationships of the more primitive classes of land plants (e.g. Banks 1975), though interestingly this knowledge has caused diversification of opinions rather than generating a consensus.

Taylor (1988, fig. 31) argued that *Rhynia gwynne-vaughanii* (Kidston & Lang 1917, 1920a, D. S. Edwards 1980) and *Horneophyton ('Hornea') lignieri* (Kidston & Lang 1920a, 1921a) represent the true rhyniopsids, a distinct class derived from the ancestral plexus of putatively pteridophytic cooksonioids (see also Kenrick & Crane in press). Both species are short (less than 30 cm) plants with centrarch xylem, showing clear division of labour between prostrate and vertical axes. The ramifying prostrate axes bear rhizoids (unicellular epidermal emergences), fulfilling the physiological roles played by the rootstocks of more advanced tracheophytes (anchorage and absorbance) and allowing asexual reproduction by vegetative division. The upright, dichotomous axes lack lateral appendages and terminate in homosporous elongate sporangia; these are simple and abscised in *Rhynia*, typically twice dichotomous, columellate, persistent and apically dehiscent in the cormose *Horneophyton* (Eggert 1974, El-Saadaway & Lacey 1979b). Branching is more frequent in *Rhynia*, where lack of vascular continuity into the anisotomous ultimate branches suggests adventitious growth (D. S. Edwards 1980). Abscission of such branches could have offered an additional means of vegetative reproduction.

Aglaophyton ('Rhynia') major (Kidston & Lang 1920a) resembles *R. gwynne-vaughanii* in growth habit and (superficially) in anatomy. However, D. S. Edwards (1986) demonstrated that *Aglaophyton* lacks true tracheids and should be relegated to the cooksonioid plexus. More recently, Kenrick & Crane (1990, in press) questioned the nature of the 'tracheids' of *R. gwynne-vaughanii*, thereby challenging its status as a tracheophyte. Furthermore, both *Aglaophyton* (D. S. Edwards 1986, Taylor 1988) and *Horneophyton* (Stewart 1983, Meyen 1987) have been implicated as sister-groups to the bryophytes (Gensel in press, Kenrick & Crane in press).

Asteroxylon mackiei (Kidston & Lang 1920b) has a similarly chequered taxonomic history, prompting arguments as to whether it should be regarded as (1) sister-group to, or (2) a primitive member of, the class Lycopsida (Fig. 2.5). Despite its unvascularized microphylls, the discovery of attached reniform sporangia (Lyon 1964) increased its perceived similarity to bona fide lycopsids (Chaloner & MacDonald 1980, Hueber in press). It shares the modest stature (*c.* 25 cm) and prostrate–erect growth habit of the other Rhynie species. Its skeletal trabecular cortex may indicate 'cheap' construction rather than being the adaptation to standing water envisaged by Kidston & Lang (1920b). A less well-known Rhynie plant, *Nothia aphylla*, is now regarded as possessing reproductive characters of the zosterophyllopsids but centrarch axial anatomy more typical of the rhyniopsids and cooksonioids (El-Saadaway & Lacey 1979a, Meyen 1987).

Much attention has been paid to the possible occurrence of gametophytes isomorphic with the sporophytes of the Rhynie cooksonioids and rhyniopsids; these would elucidate the life-cycles of early land plants. Reports of gametophytes of *Rhynia* (Pant 1962, Lemoigne 1968) were discounted by D. S. Edwards (1980), but *Lyonophyton rhyniensis* remains a candidate for a gametophyte, either of *Aglaophyton* (Stewart 1983) or of *Horneophyton* (Remy & Remy 1980, Meyen 1987). Remy & Hass (1991) recently resurrected and supplemented some of these hypotheses with new data.

Recognition of evidence for plant–animal interactions has re-invigorated studies of other organisms at Rhynie, especially arthropods. Wounds penetrating the phloem of *R. gwynne-vaughanii* axes (Kidston & Lang 1921b, Knoll 1985a) have been attributed to sap-sucking mites such as *Protacarus cranii* (Kevan *et al.* 1975). Other arthropods present include characteristically soil-dwelling herbivorous springtails (*Rhyniella praecursor*) and the large, putatively carnivorous arachnid *Palaeocteniza crassipes* (Rolfe 1980, Scott 1984). Arguments that trigonotarbid arachnids (*Palaeocharinus* sp., cf. *Palaeocharinoides* sp.) found in *Rhynia* sporangia may have consumed and thereby dispersed the spores (Kevan *et al.* 1975) were challenged by Rolfe (1980) on several grounds; some of the trigonotarbids are rotting carcasses penetrated by fungal hyphae. Even more worrying is Crowson's (1970) assertion that the mite and the springtail are later contaminants. Thus, the Rhynie Chert provides evidence, albeit equivocal, of food chain development and of both parasitic (sap-sucking) and possibly mutualistic (sporophagy) plant–animal interactions (Rolfe 1985, Labandeira & Beall 1990).

2.6 OXROAD BAY, SCOTLAND (LOWER CARBONIFEROUS, PETRIFACTION/ ADPRESSION)

Northern Britain possesses by far the greatest global concentration of localities yielding anatomically preserved Dinantian plants. Palaeobotanists attracted to these floras included (in chronological order) W. C. Williamson, D. H. Scott, R. Kidston, W. T. Gordon, J. Walton and A. G. Long. Their work generated an exceptional database of rigorous form-species descriptions, though this was only rarely exploited in attempts to reconstruct whole plants (e.g. Walton 1935, Long 1979) or to interpret the geological context of the assemblages (e.g. Gordon 1909, 1938). Recently, interest has been renewed in the depositional environments of the fossils and in palaeoecologically oriented floristic analysis (e.g. Scott *et al.* 1984, Scott & Rex 1987). Although the region has yielded many fluvio-lacustrine assemblages, the most detailed studies have been performed on volcanigenic terrains: these are rhyolitic at Esnost, France (Rex 1986b) but basaltic in south-east Scotland at the Pettycur–Kingswood complex (Scott *et al.* 1986, Rex & Scott 1987) and at Oxroad Bay (Bateman 1988, Bateman & Rothwell 1990, Bateman & Scott 1990). Studies of all three areas shared the same basic approach: geological survey and mapping, detailed lithological logging of plant-bearing exposures, stained petrographic thin-section studies elucidating modes of preservation, and qualitative lists of petrifaction form-species based on thin-section and acetate peel studies (together with lists of adpression form-species if present). Unusually cautious plant identifications entailed liberal use of open nomenclature, and plant-bearing strata were dated by comparing associated miospore assemblages with the well-established regional palynozonation (e.g. Clayton *et al.* 1977, Scott *et al.* 1984).

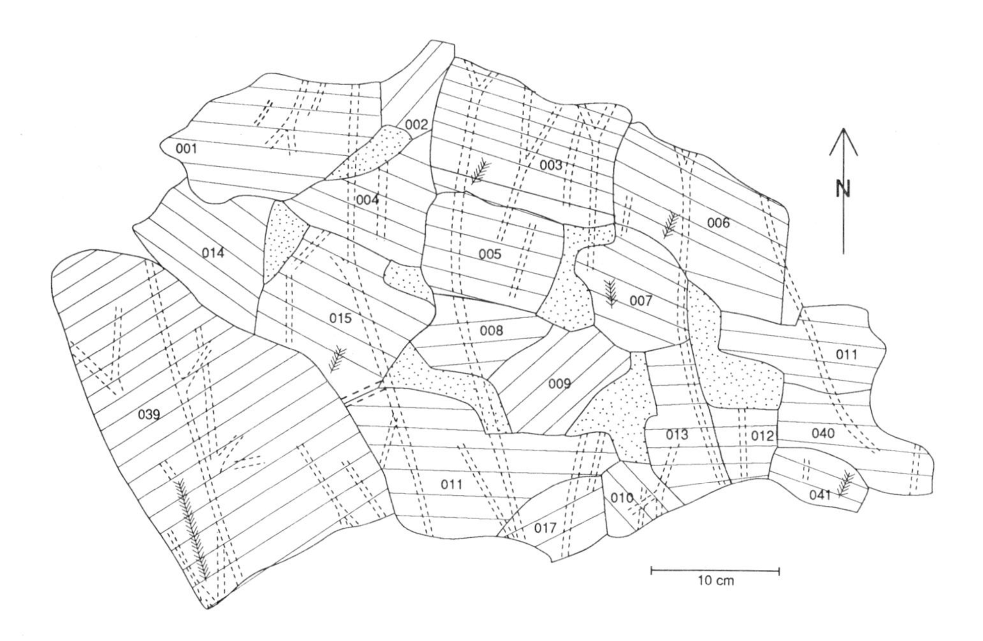

Fig. 2.7 — Reconstructed limestone horizon (Oxroad Bay, exposure D, horizon 1.23) showing individual numbered blocks and the saw-cuts that allowed the passage through the matrix of *Oxroadia gracilis* axes (dashed lines) and cones (chevrons) to be traced. Slight dextral curvature probably reflects structural deformation (Adapted from Bateman & Scott 1990, Fig. 15(a)).

The Oxroad Bay project is the most recent and broadly based investigation. Miospore dating of the strata (Scott *et al*. 1984) indicates the late Tournaisian (Tn3: *c*. 353 Ma following Leeder 1988). The locality is situated along the coast of East Lothian, close to the south-east margin of the Scottish Midland Valley. This region of the southern margin of Laurasia was tectonically active during the Dinantian (e.g. Leeder 1987). Sedimentation of the dominantly non-marine Calciferous Sandstone Group in the Oxroad Bay area was strongly influenced by the development of a series of tuff-ring ash-cones, notably the adjacent Tantallon diatreme (Leys 1982). Palaeolatitude is contentious; suggestions range from *c*. 15°S (Johnson & Tarling 1985) to *c*. 10°N (Scotese 1986). This uncertainty has weakened palaeoclimatic interpretations. Moreover, the general aridity claimed for the region (e.g. Raymond 1985, Raymond *et al*. 1989) may have been ameliorated and climate rendered exceptionally unstable in the Oxroad Bay area by proximity to the proto-North Sea and by periodic eruptive seeding of the atmosphere with ash particles. Climatic evidence from the plants themselves is highly equivocal (Bateman & Scott 1990).

Several fossil plant assemblages span *c*. 40 m of pyroclastic-rich sediments; all experienced at least short-distance transport and most are calcareously permineralized (only horizons containing more than 30% carbonate yielded petrifactions: Fig. 2.9). Eight plant-bearing exposures occur at Oxroad Bay and two at Castleton Bay, on the opposite side of the Tantallon diatreme (Scott & Galtier 1988, Galtier & Scott 1990). Although correlation was hampered by severe volcanically induced tectonism and geochemical overprinting, Bateman & Scott (1990) resolved the fossil plant assemblages into five putative palaeocatenas (fossil landscapes). At least one assemblage occurs in each of the four major depositional units (1–4: Table 2.2). Adpression assemblages were used primarily as taxonomic adjuncts to the petrifactions.

The Oxroad Bay study included two novel features. Firstly, extensive areas of permineralized horizons were disaggregated in the field and reconstructed in the laboratory to trace the passage of plant parts through the matrix (Fig. 2.7). Intended primarily to aid whole-plant reconstruction, this also allowed clast orientation measurements (Fig. 2.8; e.g. Briggs 1977). Secondly, inductively coupled plasma (ICP) analysis (e.g. Dahlquist & Knoll 1978) generated large amounts of geochemical data for interpretation via multivariate ordinations (Fig. 2.9). These facilitated both the stratigraphical correlation of *in situ* plant-bearing horizons and the provenancing of eroded loose blocks bearing important plant remains (labelled 'LB' on Fig. 2.9). The geochemical data also elucidated plant preservation (Bateman & Scott 1990).

Much of the effort expended at Oxroad Bay has been directed towards reconstructing whole-plant species from the 43 anatomically preserved and 19 adpressed form-species occurring at the locality (Bateman & Rothwell 1990, tables 2–3). Current estimates indicate that 17–18 whole-plant species of five classes are preserved in the Oxroad Bay area (Table 2.2). Some have been fully reconstructed (the two *Oxroadia* species: Bateman 1988, Bateman *et al*. in press; *Protocalamites longii*: Bateman in press), others are in the process of complete or partial reconstruction (*Cladoxylon waltonii*: Bateman in prep.; *Bilignea* cf. *solida*: Bateman & Long in prep.; several other pteridosperms: Rothwell and co-workers in prep.), yet others

Table 2.2 — Stratigraphic distibution of anatomically preserved, putative whole-plant species in the Oxroad Bay area of south-east Scotland. Numbers preceding binomials indicate the current status of whole-plant reconstructions: 1, poorly known; 2, partially reconstructed; 3, fully reconstructed; p, potential status pending further work; parenthetic numbers indicate the status of species reconstructed using specimens from other localities. Abundance of putative whole-plant species: *, frequent; +, rare (not assessed for the inaccessible and therefore under-recorded exposure E of unit 4). Based on data in Bateman & Rothwell (1990) and Bateman & Scott (1990) for Oxroad Bay (OB), and Scott & Galtier (1988) and Galtier & Scott (1990) for Castleton Bay (CB).

Depositional unit/landscape		1	2	3	4a	4b
Locality		CB	OB	OB	OB	OB
Petrifaction-bearing exposure(s)		—	C	A	D,G,?H	E
No. of petrifaction-bearing horizons		?2	1	4	16	1+
No. of blocks studied		?	106	254	211	6
Supplementary adpressions		–	+	–	(+)	–
Lycopsida						
1	New genus A		+	+	+	
3	*Oxroadia gracilis*[1]				*	+
3	*Oxroadia* sp. nov.		*			
p2(2)	'*Lepidodendron*' *calamopsoides*[2]	*				
Shenopsida						
3	*Protocalamites longii*		*			
Pteropsida						
p2	*Cladoxylon* cf. *waltonii* +	*		+		
p2(2)	*Stauropteris* cf. *berwickensis*				+	
1(2)	*Protoclepsydropsis kidstonii*	+				
?Progymnospermopsida						
1	*Protopitys scotica*	+				
Gymnospermopsida: pteridospermales						
p2(2)	*Stenomyelon tuedianum*	*				
p3	*Buteoxylon*/*Triradioxylon* spp.[3]		*[4]	*	*	+[4]
p3	*Tetrastichia bupatides*			*		
p3	*Bilignea* cf. *solida*		*	*	+	
2	*Eristophyton beinertianum*	+	*[4]	*	*[4]	?+[4]
2	*Eristophyton waltonii*	*				
p2	New genus A (Stem H)[5]			*		
1	New genus B (Stem I)[5]		+			
Min. total no. of whole-plant species		8	8	6	7	3[6]
Min. no. of pteridospermous whole-plant species		3	4	5	3	2
No. of ovule-species		2	9	5	7	1

[1]A record of *Oxroadia gracilis* from Castleton Bay (Scott & Galtier 1988) is omitted; the single rhizomorph in question is indistinguishable from that of *O.* sp. nov.

[2]Probably more closely related to *Anabathra* (syn. *Paralycopodites*) than to *Lepidodendron s.s.* (Bateman *et al.* in press).

[3]Although ostensibly including four stem-genera (*Buteoxylon, Triradioxylon, Calathopteris, Oxroadopteris*), this group probably contains only one or two whole-plant species (Rothwell in Bateman & Rothwell 1990: 154).

[4] Axes absent; evidence consists largely of petioles known to co-occur with particular stem-species in other exposures.

[5]Parenthetic designations follow Bateman & Rothwell (1990: 137). Stem H was formerly assigned to *Tristichia ovensii*.

[6]Value undoubtedly an underestimate due to small sample size.

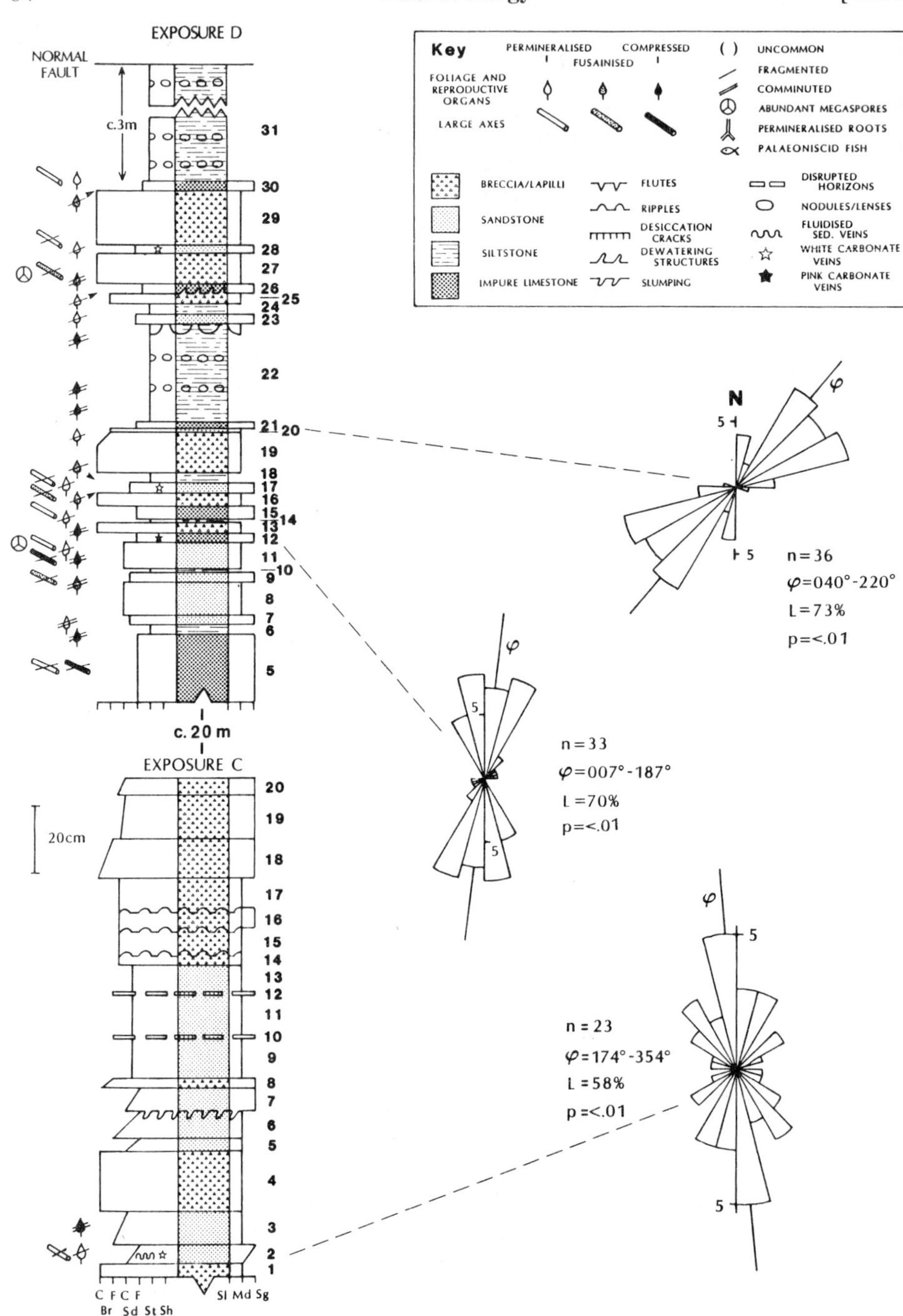

Fig. 2.8 — Lithological logs of exposures C and D at Oxroad Bay, together with orientation diagrams for petrified plant fragments in three horizons. Left-hand column of log indicates texture, right-hand column indicates induration. Rose diagrams show orientations of larger (greater than 2 cm) plant fragments (n=number of measurements; φ=resultant vector, L=vector magnitude, p=statistical significance via Rayleigh test); rose diagram for horizon D12 is taken from Fig. 4. (Adapted from Bateman & Scott 1990, Figs 8, 11, 13–15.)

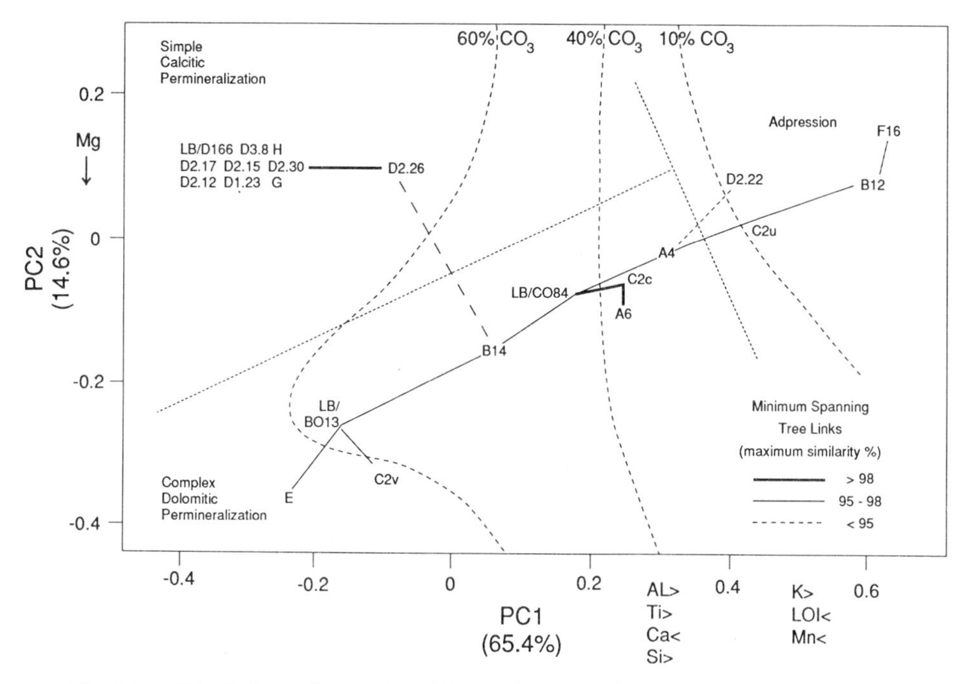

Fig. 2.9 — Principal coordinates plot with superimposed minimum spanning tree for ICP-determined major elements in plant-bearing horizons at Oxroad Bay. Data points are 19 plant-bearing horizons from eight exposures (A–H), together with three eroded loose blocks (LB). Vector strengths, and elements contributing to the vectors, are also shown. Dashed lines are carbonate content isopleths; dotted lines delimit modes of plant preservation. All minimum spanning tree links within the limestone group exceed 98% similarity. (Adapted from Bateman & Scott 1990, Fig. 18, Table 8.)

are so poorly represented that they will remain enigmatic. Rather than discussing in detail the morphology, palaeobiology and phylogenetic implications of these plants, I shall briefly (and qualitatively) describe each floral assemblage and associated palaeocatena to illustrate how our current (albeit tentative) understanding of the ecosystems stems from integration of many ostensibly disparate sources of palaeobotanical and palaeoenvironmental data.

The earliest depositional unit (1) occurs at both Oxroad Bay and Castleton Bay, but contains plant fossils only at the latter. Volcanicity was subdued, being confined to small cinder cones (red-group maars) that had little effect on local topography or sedimentation. Permineralization was simple, calcitic and probably slow. Scott & Galtier (1988) interpreted the enclosing fine sands and silts as crevasse-splay and overbank deposits, laid down by braided rivers crossing a lowland flood plain.

The nominal eight whole-plant species in unit 1 include the presumed small tree-pteridospermalean *Eristophyton beinertianum*, the only species occurring in all five assemblages. Here it is accompanied by the congeneric species *E. waltonii*. The other two dominant elements, the (also presumed) small tree-lycopsid '*Lepidodendron*' (probably *Anabathra*) *calamopsoides* and shrubby pteridospermalean *Steno-*

myelon tuedianum, co-occur in Tournaisian fluvio-lacustrine sediments elsewhere in SE Scotland (Scott & Rex 1987) but are absent from the overlying sediments at Oxroad Bay, suggesting that they were unable to survive in the succeeding, volcanically modified landscapes. The dominant species in unit 1 are represented by some large axes and rootstocks, perhaps indicating short-distance transport.

Although most assemblages in unit 2 are highly disarticulated and adpressed, fortuitous injection of fluidized sediment veins allowed rapid and complex dolomitic permineralization of a small lens that contains some well-articulated specimens. Elsewhere on this palaeocatena, subaqueous ionic concentration occasionally and gradually formed secondary cementstones. A range of sedimentary features and the presence of intact palaeoniscoid fish suggest that the plants were deposited in a shallow, periodically emergent lake that probably separated two juvenile tuff-rings. Any directional flow oriented only the smaller plant fragments (Fig. 2.8). Ovule-species in particular are heterogeneously distributed within the lens. Enclosing sediments contain greater proportions of reworked pyroclastic material, indicating the emergence of at least some of the more prominent green-group tuff-rings.

The eight whole-plant species in unit 2 include only two species found in unit 1: *E. beinertianum* and the short, shrubby cladoxylopsid *Cladoxylon* cf. *waltonii* (Table 2.1). Other dominant elements are unique to this unit: the shrubby sphenopsid *Protocalamites longii* and small, semi-prostrate (but nonetheless determinate and wood-producing) lycopsid *Oxroadia* sp. nov. All three species are short (below 1 m) and have swollen, woody rootstocks and/or stems — obvious assets in unstable soils experiencing at least periodic water deficits. The low degree of disarticulation and presence of rootstocks suggest origination from a local (possibly riparian) community. Although severely disarticulated, three of the associated pteridosperms are frequent, but the fourth (new genus B) occurs only as a single pyrofusain fragment. *Eristophyton* and *Bilignea* appear capable of developing into trees at least several metres high, whereas members of the enigmatic *Buteoxylon/Triradioxylon* complex may not have exceeded shrub size (Bateman & Rothwell 1990).

Unit 3 includes the classic cliff-section assemblage of Gordon (1938); a large-scale, trough- and cross-stratified jumble of coarse, greyish-green pyroclastic debris that was deposited by a series of base surges and/or mass flows, presumably induced by the nearby Tantallon vent. These events actively interred plants of varying degrees of disarticulation in a wide range of orientations. The complex polyphase mineralization and partial devolatilization imply slow, passive permineralization. Floristic variations evident among different horizons have not been studied in detail; finer-grained lenses rich in ovules may reflect aeolian winnowing. Presumably, the plants of unit 3 grew higher on the slopes of the Tantallon tuff-ring, were violently decapitated (rootstocks are absent), and transported over a short distance either prior to or immediately after burial.

At least six whole-plant species occur in unit 3 (Table 2.1). The new, ostensibly herbaceous lycopsid genus is represented only by a single axial fragment. The remainder are pteridospermaleans, apparently a mixture of shrubs and small trees. Three species that occurred on the preceding landscape (*E. beinertianum, B.* cf. *solida, Buteoxylon/Triradioxylon*) are joined by two shrubby, apparently *r*-selected pteridospermaleans (*Tetrastichia bupatides*, the new genus A) that are confined to unit 3. All five occur in quantity, and smaller axes in particular have retained petioles. It is tempting to interpret the absence of free-sporing plants as indicating at least temporary restriction of free water.

Considerably thicker than unit 3, unit 4 also consists largely of volcanic ash. Much of this is believed to have been reworked, suggesting extensive and prolonged landscape disturbance. It contains two stratigraphically and palaeoenvironmentally distinct depositional environments that include plant megafossil assemblages. The earlier (4a) is a post-depositionally faulted and folded pseudochannel; this exposes a thin (*c*. 4 m) but sedimentologically complex sequence of intercalated coarse tuffs, finer-textured and less-cemented reworked pyroclastics, and 16 calcitic limestone horizons that differ in purity and petrified plant content (Fig. 2.8). The simple micritic calcite permineralization contrasts with the complex, dolomite-rich permineralization that characterizes the other assemblages (Fig. 2.9); the latter reflects the mobile element content of the abundant tuffs. These limestones may represent biogenic precipitation of calcite in a thermally stratified eutrophic lake by algal blooms that flocculated and enveloped plant megafossils deposited on the lake floor (cf. Loftus & Greensmith 1988). This passive depositional scenario contrasts with the directional (if fluctuating) flow indicated by orientation studies of axes in two of the limestones (Figs 2.7–2.8). Two horizons near the base of the sequence are poorly developed palaeosols that contain permineralized pteridospermalean roots, indicating brief emergence and colonization. In addition, the putative arthropod coprolites that occur in all the plant assemblages are here associated with arthropod cuticle; some of the coprolites demonstrate species-specific herbivory (Bateman & Scott 1990).

The presence of 16 petrifaction-bearing horizons offers an unusually good opportunity to examine association/dissociation patterns of form-species (Bateman & Rothwell 1990, table 4); these data can be integrated with preservational and palaeoenvironmental information to infer the whole-plant species composition of communities contributing to these transported assemblages. Together, the 16 horizons yielded seven whole-plant species (Table 2.1). Rare components include the only Oxroad Bay specimens of the small, probably prostrate pteropsid *Stauropteris* cf. *berwickensis* and fusainized fragments of the cladoxylopsid *Cladoxylon* cf. *waltonii*. Pteridospermaleans are less diverse than in unit 3; all three species in unit 4a also occur in units 2 and 3 (here they are represented largely by abscisable organs; axes are rare). The most notable species first occurring in unit 4 is the pseudoherbaceous lycopsid *Oxroadia gracilis*, which is represented by substantial quantities of most component organs (only rhizomorphs are rare), and is well-articulated compared with other co-occurring species. *Oxroadia gracilis* dominates several horizons in the central portion of the succession, including some thin concentrations of pyrofusain; these suggest that wildfires (perhaps volcanically induced) swept through dense monotypic stands of this lycopsid, which appears well adapted to the highly disturbed landscape of unit 4. The putative pteridospermalean scrub community of *Eristophyton–Bilignea–Buteoxylon/Triradioxylon* persisted on this landscape.

Limited study of an inaccessible exposure higher in the unit (4b) yielded only the three dominant species of unit 4a (*O. gracilis, E. beinertianum, Buteoxylon/*

Triradioxylon). Lacustrine deposition is again envisaged, though permineralization returned to the dominantly slow, dolomitic mode that characterizes units 1–3 (Fig. 2.9); partial devolatilization is reminiscent of that in unit 3.

Reviewing the locality as a whole, some quantitatively rare floristic elements apparently represent true ecological rarity (e.g. lycopsid new genus A), whereas others probably represent fortuitous preservation of highly allochthonous material (e.g. pteridosperm new genus B). Some of the rarer species were probably genuinely herbaceous (lycopsid new genus A, *Stauropteris, Protoclepsydropsis*). However, all of the ecological dominants of all of the communities were capable of secondary thickening, though wood was often thin (e.g. *Protocalamites*) or highly localized within the plant body (e.g. *Oxroadia*). Of the two pairs of congeneric species present, the two *Eristophyton* species co-occur in only one assemblage and the two *Oxroadia* species in none (Table 2.2), despite the high potential for taphonomic mixing. This suggests that ecological preferences differed in detail among closely related species. Observed species 'turnover' rates among units are 30–60%, with the greatest change, between units 1 and 2, coinciding with the onset of widespread volcanism in the vicinity. Pairwise compositional comparisons of the assemblages show that units 2 and 3 are most similar in ecological dominants (sharing three species), but that units 2 and 4 are most similar in overall diversity (sharing five species). Only *Cladoxylon* shows a stratigraphical 'gap' (in unit 3); this could be interpreted as indicating local extirpation followed by recolonization from nearby populations.

2.7 DUCKMANTIAN COAL MEASURES, BRITAIN (UPPER CARBONIFEROUS, ADPRESSION/SPORE)

The Upper Carboniferous coal basins of Laurasia have been the cradle of Palaeozoic plant palaeoecology. Their attraction reflects the relative abundance of plant fossils (adpressed and petrified) and of industrially created exposures, together with the use of plant fossils as the primary stratigraphical tools (Cleal, this volume). Early speculations on coal-swamp ecosystems by taxonomically oriented palaeobotanists (e.g. Potonié 1899) were superseded by semi-quantitative (Davies 1929, Drägert 1964) and then fully quantitative (Oshurkova 1974, Scott 1977) studies of adpressions. Here, I have chosen to illustrate palaeoecological studies using Scott's (1977, 1978, 1979) pioneering work, which focused on British Duckmantian (*c.* 310 Ma) Coal Measures, and counterpoint the Westphalian D American coal-ball studies described in section 2.8. In Duckmantian times, Britain was situated close to the eastern margin of the Laurasia palaeocontinent, at a palaeolatitude of 0.5°N (e.g. Scotese 1986). Climate was equatorial tropical, though evidence from American coals suggests a period of relative dryness during the early (Phillips & Peppers 1984) or middle (Winston 1990) Duckmantian. The numerous coal-forming basins of the British Westphalian largely reflected tectonically induced block subsidence earlier in the Carboniferous (cf. Bott 1987, Grayson & Oldham 1987, Leeder 1987); occasional sedimentary influxes were of dominantly northern provenance.

Scott (1977, 1978, 1979) gathered quantitative data on adpression assemblages from 17 Coal Measures profiles in northern England and southern Scotland. The logged sections were partitioned into horizons bounded by discontinuities in lithology and/or sedimentary structures. Within each horizon, a bedding plane was cleared and the exposed adpression assemblage was quantified using a 0.5-m^2 quadrat with *c.* 100 random sample points (Scott & Collinson 1983, cf. Spicer 1988). Lateral replicates were not attempted, and in each case the majority of points sampled only bare rock. Primary data were published for only one of the 17 profiles studied: Annbank, 45 km south of Glasgow in the Ayrshire coalfield (Scott 1977). Here, megafossil data were obtained from a 1.6-m-thick profile of roof-shales immediately overlying the thin (12-cm) Top Ell Coal (uppermost Duckmantian). This spanned three upward-fining beds. The lower two were silty, often plane-laminated, and interpreted as lacustrine delta or crevasse-splay deposits. The uppermost was richer in sand, contained relatively poorly preserved adpressions, and was interpreted as a more proximal alluvial deposit.

Of the 23, 3–16-cm-thick horizons, 18 yielded adpression assemblages (Fig. 2.10). Percentage cover data were converted to the ten-level Domin Scale. Allowing for the problems inherent in adpression taxonomy, the 30 form-species recorded in the profile could represent as few as 12 whole-plant species; of these, only six exceeded 5% cover in any horizon. The sequence is dominated by the foliage of *Alethopteris* and at least two species of *Neuropteris* (all putative pteridosperms) and the cordaite *Cordaites*, with subordinate *Mariopteris* (another putative pteridosperm) and at least one calamite. Except for two horizons (3b, 4f) containing only calamite roots, all of the plant fossils were undoubtedly transported; taphonomic sorting and/or degradation may explain the tendency for monospecific dominance in the coarser horizons. Otherwise, the data lack obvious stratigraphical trends (Fig. 2.10). This seed-plant-dominated flora contrasts greatly with a miospore assemblage extracted from the underlying coal, which was dominated by lycopsids with subordinate calamites (Scott 1977).

Scott presented the remainder of his extensive database semi-quantitatively in papers summarizing five Westphalian sections from the Wakefield area of the North Yorkshire coalfield (Scott 1978) and, subsequently, all 17 study sections distributed throughout northern Britain (Scott 1979). The adpression assemblages were described using scaled frequency of occurrence among horizons and average cover within horizons, and were placed in the context of idealized lithological logs translated into idealized lithofacies (Fig. 2.11). Scott (1979) recognized four major depositional environments (lake, deltafill, floodplain, saline) encompassing 13 lithofacies, though plant megafossil data were restricted to the first three 'megafacies' and to nine of the facies.

Lacustrine environments included prodelta shales rich in lycopsids (notably '*Lepidodendron*' *s.l.*) with some calamites, and river distributary mouthbar sequences where monotypic calamite stands (some found *in situ*) prevailed. Of the three deltafill facies, two (proximal distributary mouthbars and interdistributary lakes) contained highly transported, largely unidentifiable plant debris. Marginal lake and delta-top deposits yielded diverse assemblages that included calamites and a range of fern-like foliage (*Neuropteris, Mariopteris, Alethopteris, Sphenopteris, Diplotmema*), occasionally with the arboreous lycopsid *Bothrodendron* and cordaitaleans. Bona fide floodplain sediments and crevasse-splays, arguably the most

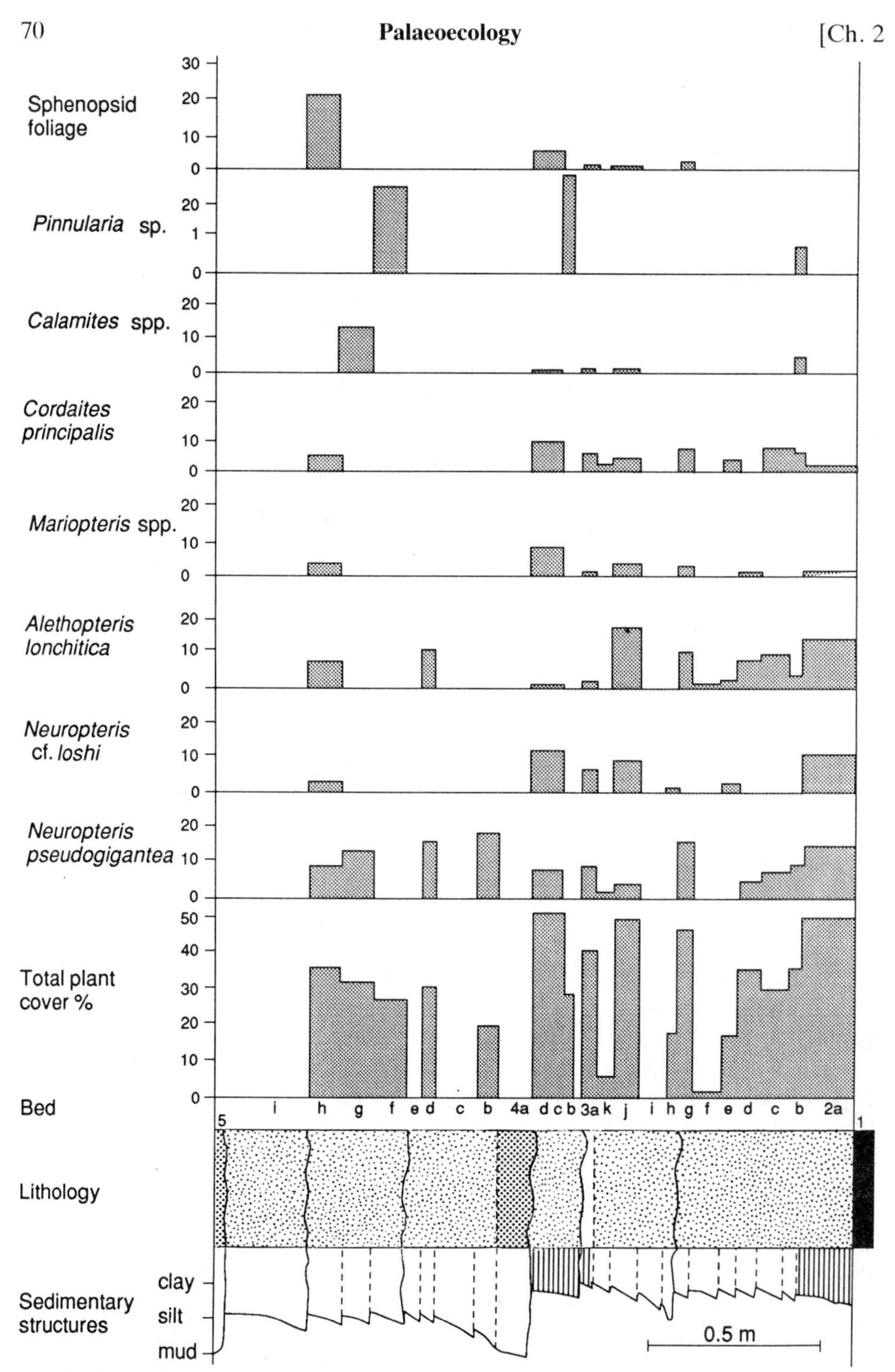

Fig. 2.10 — Stratigraphic distribution of adpression form-taxa through a section at Annbank. Histograms represent percentage cover estimates based on point quadrat analysis. Data were not presented for horizons 2i, 3b, 4a, 4c, 4e, and 4i. (Redrawn from Scott 1977, Fig. 7.)

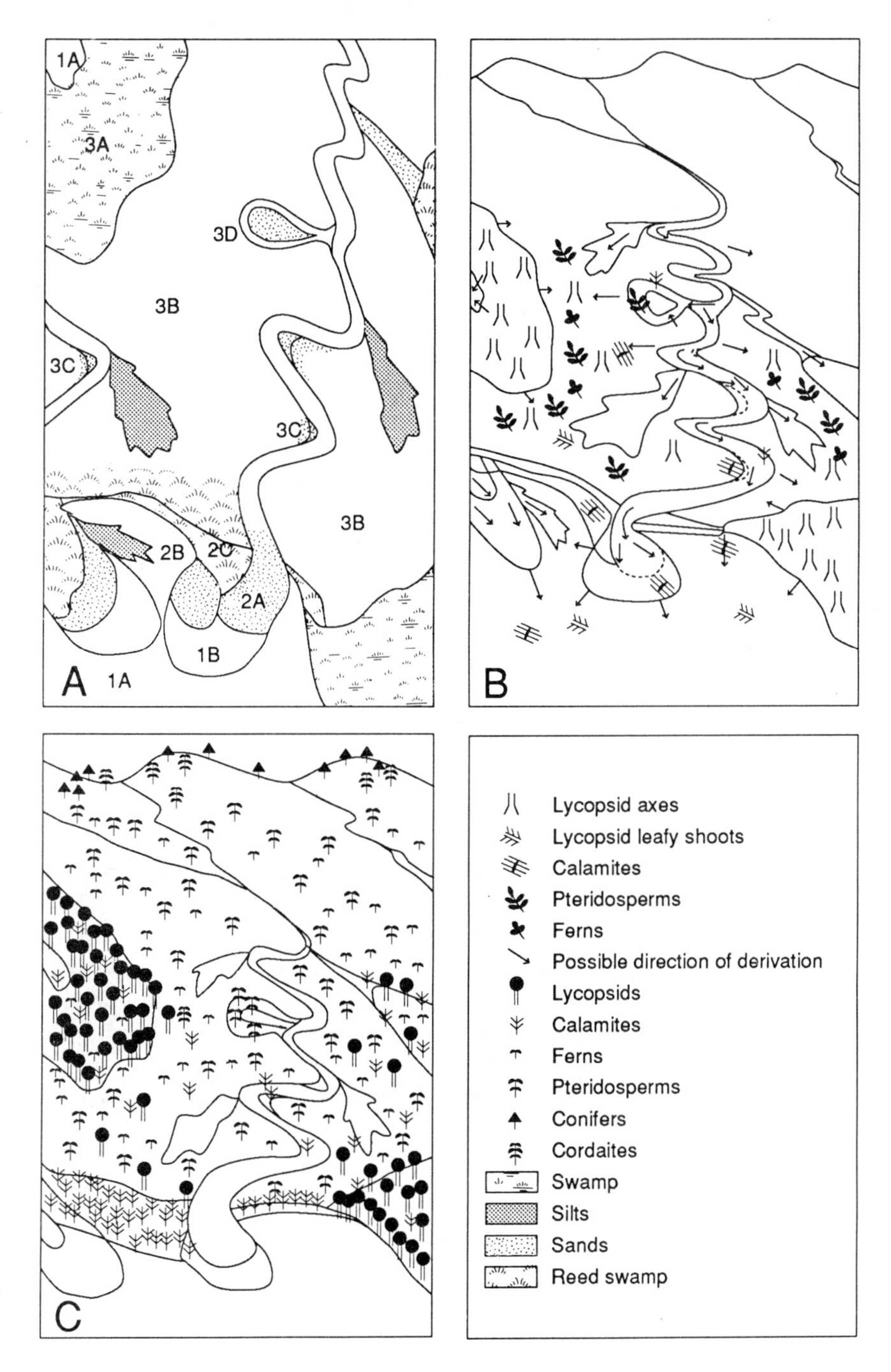

Fig. 2.11 — A lithofacies-based approach to visualizing fossil plant communities. (A) Lithofacies representing a hypothetical landscape in the Duckmantian B of Britain (1, lake; 2, deltafill; 3, floodplain). (B) Possible taphonomic trajectories. (C) Reconstruction of plant community mosaic based on integration of (A) and (B). (Redrawn from Scott 1979, Fig. 6.)

diverse lithofacies, yielded the most diverse fossil plant assemblages (including the roof-shale assemblages described above). Fern-like foliage (especially *Neuropteris*) and calamites dominated, while subordinate elements included sphenophyllaleans, cordaitaleans, and a range of arboreous lycopsids. More depauperate lithofacies are point-bars (occasional *Neuropteris*, calamites, and arboreous lycopsids) and channel-fills (fern-like foliage). Scott (1978, Figs 15–17) relied on spore assemblages, sampled in 1-cm-thick intervals spaced 10 cm apart, to indicate the floral composition of the coalified peats; these suggested lycopsid dominance or lycopsid–calamite codominance throughout.

Scott (1978, 1979) identified repeated cycles of lithofacies (typically lake>delta-fill>floodplain) through the Coal Measures, and noted that the relative proportions of these facies differed among penecontemporaneous basins. In general, there is an upward increase in alluvial facies through the Coal Measures; previous workers under-estimated floodplain deposition and over-estimated the importance of marine transgressions. Peat accumulation was variously terminated by lake or channel formation, or by marine incursion. Scott regarded coal-swamp peat floras as autochthonous, but recognized the dominantly transported nature of fossil plant assemblages from the associated clastic deposits; he also noted the possible incorporation of presumed extra-basinal ('upland') species such as coniferaleans and some cordaitalean species. Thus, by integrating biotic data with detailed information on bed geometry, lithology and sedimentary structures, the environments of deposition of the plant fossils were successfully reconstructed, though for many of the species present these were not the environments of growth (see for instance comments by Gastaldo 1985b).

2.8 HERRIN COAL, ILLINOIS BASIN, USA (UPPER CARBONIFEROUS, PETRIFACTION)

Coal balls are arguably the epitome of Palaeozoic palaeobotany. They attract the palaeobiologist because of their often exquisite anatomical preservation and relative articulation of the enclosed plants, and the geologist because of their widespread occurrence in economically important coals and their potential as palaeoenvironmental indicators (e.g. Phillips 1980, Scott & Rex 1985). However, only recently have serious attempts been made to integrate the exceptional databases that have accrued on the morphology of coal-ball plants and on the geology of the enclosing coals into coherent palaeoecological and palaeoenvironmental interpretations of the coal-swamps. The most comprehensive have been performed by T. L. Phillips, W. A. DiMichele, R. A. Peppers and co-workers in the Illinois Basin, where at least 15 coal-ball-bearing coals span the Westphalian D and lower Stephanian (Phillips *et al.* 1985, fig. 1). Of these, the most productive and thoroughly investigated is the upper Westphalian D (c. 307 Ma) Herrin Coal of the upper Carbondale Formation.

This sequence of fossilized peats and intervening clastic partings formed at a palaeolatitude of 0–5°S in a wet tropical climate (e.g. Ziegler *et al.* 1981, Phillips & Peppers 1984), though the rising Appalachians may have reduced precipitation in the basin from the prevailing easterly winds. The Herrin peat-swamps developed on the extensive delta of the Michigan River, which drained obliquely westward from the northern Appalachians and prograded into a shallow intrusive lobe of the extensive ocean to the west (Phillips & DiMichele 1981, Fig. 7.1, Phillips & Peppers 1984, Fig. 14). The most striking feature on this subdued deltaic topography was the sinuous Walshville palaeochannel, whose broadly north-south-oriented course approximated the margin of the delta and may have been at least partly estuarine (Wanless *et al.* 1969, Johnson 1972). Close to the palaeochannel, the coal is thicker, richer in clastic partings, and separated from the marine Anna Shale and Brereton Limestone (which immediately overlie the Herrin Coal elsewhere) by the non-marine, putative crevasse-splay sediments of the Energy Shale. The 19 Herrin Coal mines that have yielded coal balls form a north-south-east crescent; most occur within 50 km of the palaeochannel (Phillips & DiMichele 1981, Fig. 7.2). Those studied in most detail are concentrated towards the south-east end of the crescent, east of the palaeochannel: Old Ben, AMAX, Sahara, Peabody Eagle (all south Illinois), Peabody Ken, and Camp (both north-west Kentucky).

The Herrin is the thickest (typically *c.* 1.5 m) and most extensive coal in the Illinois Basin. It approximates the acmes of species richness, wetness, and frequency of marine incursions across the swamp (Phillips & Peppers 1984, DiMichele & Phillips 1985, DiMichele *et al.* 1985, Mahaffy 1985). Coal-ball formation minimized subsequent compaction; coal-ball 'pods' reach 2 m at Sahara (Phillips & DiMichele 1981) and 4 m at Old Ben (DiMichele & Phillips 1988). Both the coal-ball pods and occasional clastic partings are generally laterally impersistent, though the possibly incursive 'Blue Band' is unusually extensive. Within pods, coal balls occur in distinct, 10–50 cm thick nodular or semi-laminar horizons ('zones' *sensu* Phillips and co-workers); these too are laterally impersistent (Phillips & DiMichele 1981, Fig. 7.4), suggesting that coal-ball formation was local at any one moment in time. Horizons may have become petrified sequentially upwards (Phillips *et al.* 1977), though more recent studies suggest synchronous mineralization of stacked horizons (DeMaris *et al.* 1983).

Sampling involved orientation and colour-coding of coal balls prior to extraction. Their plant fossil form-species content was determined by placing a grid on a representative acetate peel from the median vertical cut through each coal ball (Phillips *et al.* 1977, cf. Pryor 1988). Unit area thus recorded represents unit *biovolume* of each form-species (not biomass as is often mis-stated; this is influenced by the tissue : void ratio of the fossil organ and the density of the tissue present). In addition to taxonomic identifications, pyritized and fusainized organs were distinguished from simple calcareous petrifactions to facilitate taphonomic interpretations. A minority (typically less than 20%) of the areal cover in the coal balls could not be assigned to form-species; moreover, taxonomic attribution of most rooting organs is impossible below ordinal level. Thus, data were analysed at three levels of completeness: total petrified peat, total identifiable organs, and total identifiable non-rooting organs (the latter two submatrices were normalized to per cent after each round of omissions). The facts that (1) the tree-fern *Psaronius* possessed abundant aerial roots, and (2) some form-species represent more than one whole-plant species, necessitated limited ad hoc manipulation of the raw form-species data prior to analysis (Phillips & DiMichele 1981: 241).

Results indicate that the overall dominance–diversity pattern is typical of the Westphalian D and remarkably consistent throughout the Herrin coal-swamp. General descriptions focus on class- and order-level taxa: lycopsids are dominant, pteropsids (mostly tree-ferns) and pteridospermaleans subordinate, sphenopsids rare and cordaitaleans very rare (despite being common in coals below the Herrin and in contemporaneous ocean-marginal, mangrove-type communities to the west of the Illinois Basin). This five-part categorization seems grossly simplistic compared with the overall species diversity reported by Phillips & DiMichele (1981) from the Sahara: 68 form-species represent 44 putative whole-plant species of 29 genera. However, few of these species were sufficiently frequent to play major roles in the coal-swamp ecosystem; not surprisingly, our understanding of the taxonomy, phylogeny and palaeobiology of these species is roughly proportional to their overall abundance.

Mangrove-like cordaitaleans are of low diversity (Costanza 1983) and ecologically insignificant. Pteridospermaleans include putatively lianescent lyginopterids and the shrubby, semi-prostrate callistophyte *Callistophyton boyssetii* (Rothwell 1975), but are dominated in diversity and biovolume by larger, upright medullosans, notably *Medullosa* species (Delevoryas 1955, Phillips 1981, DiMichele *et al.* 1985). The taxonomically diverse pteropsids include rhizomatous coenopteridaleans and zygopteridaleans, but again only one genus contributes appreciably to biovolume: the marattialean *Psaronius*, a peat-swamp specialist. Its several species vary in size but share the same mode of upright growth supported by aerial roots (Millay 1979, Lesnikowska 1989). Sphenopsids include a few herbaceous species of *Sphenophyllum* (Batenburg 1982); overall, these are subordinate in biovolume to the larger, shrubby calamites such as *Arthropitys* (Good 1975). Small-bodied lycopsids include the semi-prostrate *Paurodendron* and upright but unbranched *Chaloneria* (Pigg & Rothwell 1983). Again, these are subordinate, both taxonomically and ecologically, to arboreous relatives. Bizarre tree-lycopsids dominated most Westphalian coal-swamp communities, and are the best understood whole-plant species; rigorous reconstructions (e.g. DiMichele 1985, *q.v.* for earlier references) have permitted detailed phylogenetic (Bateman *et al.* in press) and autecological (DiMichele & Phillips 1985, Phillips & DiMichele in press) interpretations.

These highly specialized lycopsids are characterized by determinate modular growth from a centralized rhizomorphic rootstock. The enormous vegetative primary body generated a mature plant that was both arboreous (large body, upright growth) and arborescent (wood-producing), though it depended primarily on periderm ('bark') for structural support; indeed, periderm and associated cortical tissues dominate most Herrin assemblages (DiMichele & Phillips 1988). The narrow, short-lived leaves cast little shade, the plant relying on photosynthetic tissues diffused among many organs (possibly even the abundant, spoke-like rootlets: Phillips & DiMichele in press).

Arboreous lycopsids of the Herrin Coal comprise seven species of six genera (Table 2.3). These were assigned by DiMichele & Phillips (1985) to three basic ecological roles: colonizers (*Anabathra*), site occupiers (*Sigillaria, Diaphorodendron*), and opportunists (*Synchysidendron, Lepidodendron, Lepidophloios*). Opportunists are characterized by cheap construction (low estimated density : vo-

Table 2.3 — Phylogeny and ecologically significant characteristics of arboreous lycopsids from the Herrin Coal (Westphalian D) of the Illinois Basin (summarizes data from DiMichele & Phillips 1985, DiMichele *et al.* 1985, Phillips & DiMichele in press, Bateman *et al.* 1992, *q.v.* for further details of the intrinsic properties listed and for explanations of recent taxonomic changes). Taxa: *Anabathra pulcherrima* (ANPU), *Sigillaria approximata* (SIAP), *Diaphorodendron scleroticum* (DISC), *Synchysidendron dicentricum* and *S. resinosum* (SYsp), *Lepidodendron hickii* (LNHI), *Lepidophloios hallii* (LSHL).

	ANPU	SIAP	DISC	SYsp	LNHI	LSHL
Intrinsic properties						
Leaf attachment (base only: distinct cushion)	B	C	C	C	C	C
Cone sexuality (bisexual: unisexual)	B	U	U	U	U	U
Dispersal unit (megaspore: megasporophyll-megasporangium complex)	M	S	S	S	S	S
Megaspores per megasporangium (multiple: single)	M	M	S	S	S	S[1]
Aerial branching[2] (lateral: negligible: terminal crown)	L	N	L	C	C	C
Construction[3] (expensive: cheap)	E	E	E	C	C	C
Relative growth rate (slow: rapid)	?S	S	S	R	R	R
Reproductive period (polycarpic: monocarpic)	P	P	P	M	M	M
Overall reproductive output (high: medium: low)	H	M	L	M	H	H
Overall ecological role (colonist: site occupier: opportunist)	C	S	S	O	O	O
Extrinsic properties						
Soil moisture (moist: wet: standing water)	M	M	W	W	?W	S
Habitat disturbance (heavy: moderate: light)	H	H	M	M	M	L
Abundance						
Range of average frequency in profiles of Herrin Coal (% total peat)	0–3	0–12	2–12	1–8	0–<1	6–49

[1]Megasporangium enclosed by enrolled alations of sporophyll.

[2]Simplistic characterization of complex growth habits. Anisotomous lateral branches develop sequentially during much of the lifetime of the individual and are usually deciduous; isotomous terminal branches only develop close to the end of the individual's life-span.

[3]Essentially energy input per unit volume, which is largely a function of the relative proportions of thick-walled cells (tracheids, sclerenchyma) to thin-walled cells (parenchyma, aerenchyma). Cheaply constructed plants have low weight : volume ratios.

lume ratios) and rapid growth that culminates in dense crown branching to facilitate high-output monocarpic reproduction. Cones are unisexual, and each single functional megaspore is dispersed within its sporangium. These cone characters also occur in the site occupiers, but these were expensively constructed and slower growing, eventually reaching heights of up to 40 m. Lower-level reproductive output was maintained throughout much of the life-history of the plant (polycarpic), cones being borne on a succession of lateral branches (these were reduced to peduncles in *Sigillaria*). The colonist *Anabathra* shared many of the vegetative characteristics and the growth pattern of the site occupiers but was smaller-bodied and thereby probably matured more rapidly. Despite being polycarpic, it maintained high reproductive output via relatively unspecialized bisexual cones that released individual megaspores as disseminules. Thus, different ecological roles reflect radically different gross morphologies (intrinsic properties). Interestingly, these cannot be predicted with confidence from evolutionary relationships; the closely related sister-genera *Diaphorodendron* and *Synchysidendron* have adopted contrasting ecological roles (Table 2.3: see also DiMichele & Bateman in press).

However, ecological roles alone did not determine the relative abundance of arboreous lycopsids in the Herrin swamp; for example, the opportunists include both the most abundant (*Lepidophloios*) and least abundant (*Lepidodendron*) coal-swamp genera. For an explanation we must turn to environmental (extrinsic) factors. *Lepidophloios*, the genus that dominated the greatest number of horizons, occupied the wettest soils and least disturbed habitats; it appears well adapted for growth and widespread propagule dispersal in standing water (Phillips 1979, Phillips & DiMichele in press). Once it had occupied a site, it relied on frequent self-replacement and its high-stress habitat to deter potential competitors. The sub-dominants *Synchysidendron* and *Diaphorodendron* preferred less saturated habitats, where disturbance was sufficient to allow their establishment but insufficient to subsequently eliminate them (presumably an especially serious threat to the slower-growing *Diaphorodendron*). *Synchysidendron resinosum* could even tolerate brackish water conditions (DiMichele *et al.* 1985). *Lepidodendron* occurs in few horizons of the Herrin Coal, always in small amounts. This also occupied soils of intermediate moisture regimes, but preferred the heavier disturbance and higher nutrient levels associated with clastic-rich environments such as levees. Finally, *Anabathra* and *Sigillaria* occur infrequently but usually in quantity, often in association with clastic partings. This indicates a preference for drier, highly disturbed habitats, where nutrients were enriched by clastic input, decay of the at least periodically undersaturated peat (indicated by poor preservation and low ratios of aerial : rooting organs: Phillips & Peppers 1984, Winston 1988), and occasional wildfires. Thus, there is clear niche partitioning among arboreous lycopsids according to their morphologically defined ecological roles and physiologically defined environmental tolerances; together, these factors discouraged co-dominance.

Armed with these autecological interpretations, we can proceed to consider the synecology of a minute portion of the Herrin swamp through the period of peat deposition. By correlating three profiles spanning a *c.* 400 m transect through Old Ben, DiMichele & Phillips (1988) generated a 3.6-m-thick aggregate profile of 29 horizons. These yielded an average of 7.3 putative whole-plant species, and 25 species in total; a relatively low figure, perhaps reflecting the proximity of the Walshville palaeochannel 12 km to the west. Again, the proportion of ecologically significant species was much smaller; only 12 species exceeded 2% of the biovolume in any of the 29 horizons and only three taxa (*Lepidophloios*, *Psaronius*, *Medullosa*) exceeded this value in more than half of the horizons. Clearly, the exceptional 'stability' of coal-swamp communities emphasized by many authors is a relative concept dependent on the scale of observation.

The abundant data generated during this study were summarized on a root-free basis by multivariate ordination (Fig. 2.12). Superficially, the ordination suggests that the swamp communities can be resolved into three end-members based on the three ecological strategies described above for the arboreous lycopsids. The horizontal axis is polarized between several horizons from the centre of the profile that are heavily dominated by *Lepidophloios* (Fig. 2.13(a)) and a smaller number of horizons from the upper part of the profile that supported more diverse communities rich in *Diaphorodendron* and/or *Synchysidendron* (Fig. 2.13(c)). These can be interpreted respectively as stable habitats highly stressed by over-abundance of free water, and similarly stable but drier, less stressed habitats (DiMichele & Phillips 1988). The vertical axis provides the third end-member: three near-basal horizons rich in *Anabathra* (Fig. 2.13(b)), interpreted as unstable opportunistic communities occupying the disturbed, clastic-rich peats that characterized the early stages of swamp development.

However, the complexities of the plant assemblages are obscured by the lack of lower order axes that are inherent in the Bray–Curtis algorithm used to generate Fig. 2.12. Firstly, a single horizon (11) is dominated by *Sigillaria*, a genus virtually absent from the other horizons; this implies an exceptional ecological interlude. Secondly, *Synchysidendron* and *Diaphorodendron* co-occur in quantity in only three of the six horizons dominated by at least one of these genera (Fig. 2.12), suggesting that their occasional co-dominance is fortuitous rather than an indication of close ecological linkage. Thirdly, many of the horizons rich in *Diaphorodendron* and/or *Synchysidendron* also contain appreciable quantities of *Psaronius*, and *Anabathra*-rich horizons tend to contain both *Psaronius* and *Medullosa*. In turn, *Medullosa*-rich assemblages often contain significant amounts of calamites.

It is these more diverse assemblages, concentrated towards the centre of the ordination, that are most difficult to interpret ecologically. For example, horizon 29 yielded eight species of frequency greater than 2%; it is dominated by three lycopsid species plus *Psaronius* (Fig. 2.13(c)). In contrast, horizon 18 yielded seven frequent species but is poor in lycopsids; it is dominated by *Medullosa* and *Psaronius* with significant quantities of the pteridospermalean *Callistophyton*, cordaitalean *Pennsylvanioxylon* and calamite *Arthropitys* (Fig. 2.13(d)). Such assemblages are consistent with patchy, low canopy communities occupying disturbed habitats rich in clastic materials (DiMichele & Phillips 1988), implying greater niche partitioning than in the other Herrin swamp communities. Indeed, the fact that horizons 11, 18 and 29 all immediately followed periods of *Lepidophloios* dominance is consistent with significant changes in the local environment of growth, but it is also consistent with significant changes in the local environment of *deposition*. This distinction is crucial; do these unusually diverse assemblages constitute unstable, transient communities, or do they merely indicate taphonomic mixing of components of more than one community? Do the unusually large amounts of fusain in horizons 11 and 18 indicate environmental drying and consequent increased vulnerability to wildfire, or taphonomic influxes of burnt material from extra-basinal sources?

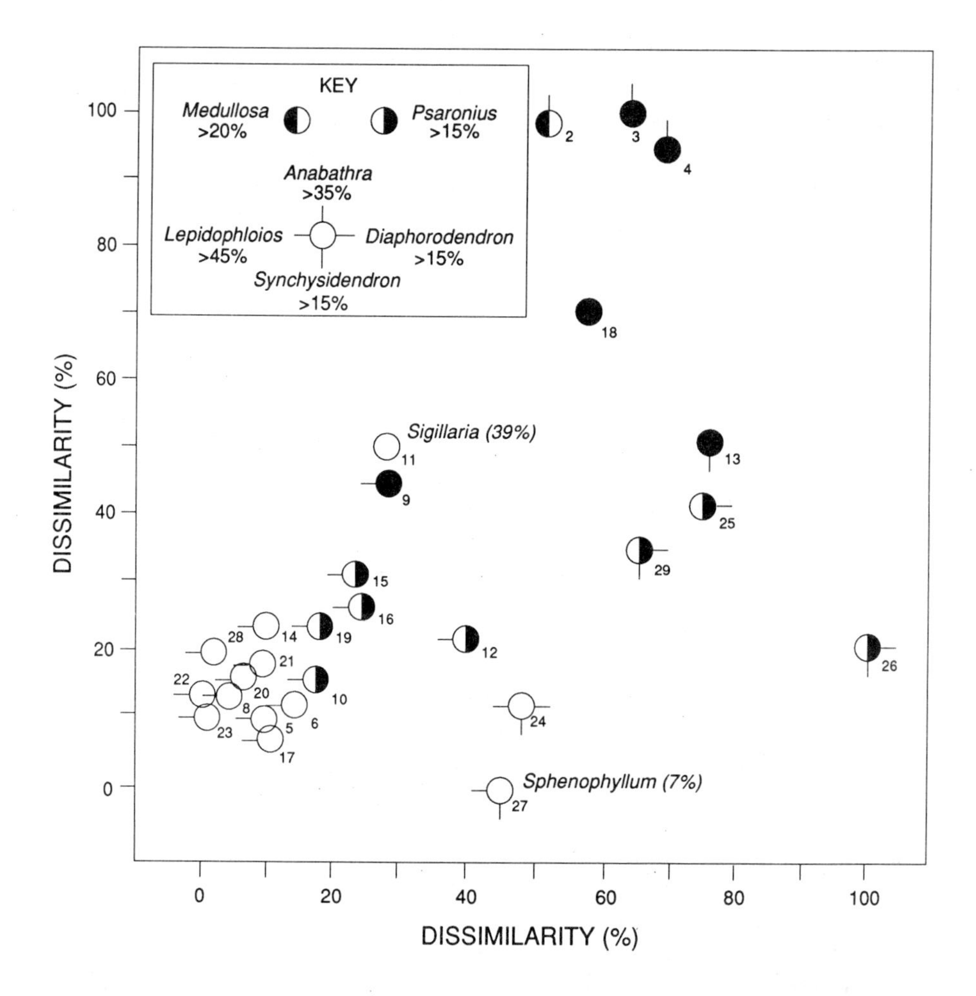

Fig. 2.12 — Indirect polar ordination of root-free biovolume data for petrifaction species from the Herrin Coal. Numbered data-points are coal-ball horizons from a composite vertical section at Old Ben No. 24 Mine; data for horizons 1 and 7 were insufficient to warrant inclusion. Symbols indicate dominant and sub-dominant taxa. Dissimilarity values used in the ordination are the negatives of similarity values calculated using an unspecified coefficient. The two axes are scaled according to end-points (horizons) selected by the analyst, with the remaining points positioned according to Bray–Curtis distances (Digby & Kempton 1987). This method summarizes all of the data and emphasizes dominant taxa, at the expense of decreases in objectivity and the ability to resolve large numbers of communities. (Adapted from DiMichele & Phillips 1988, Fig. 10.)

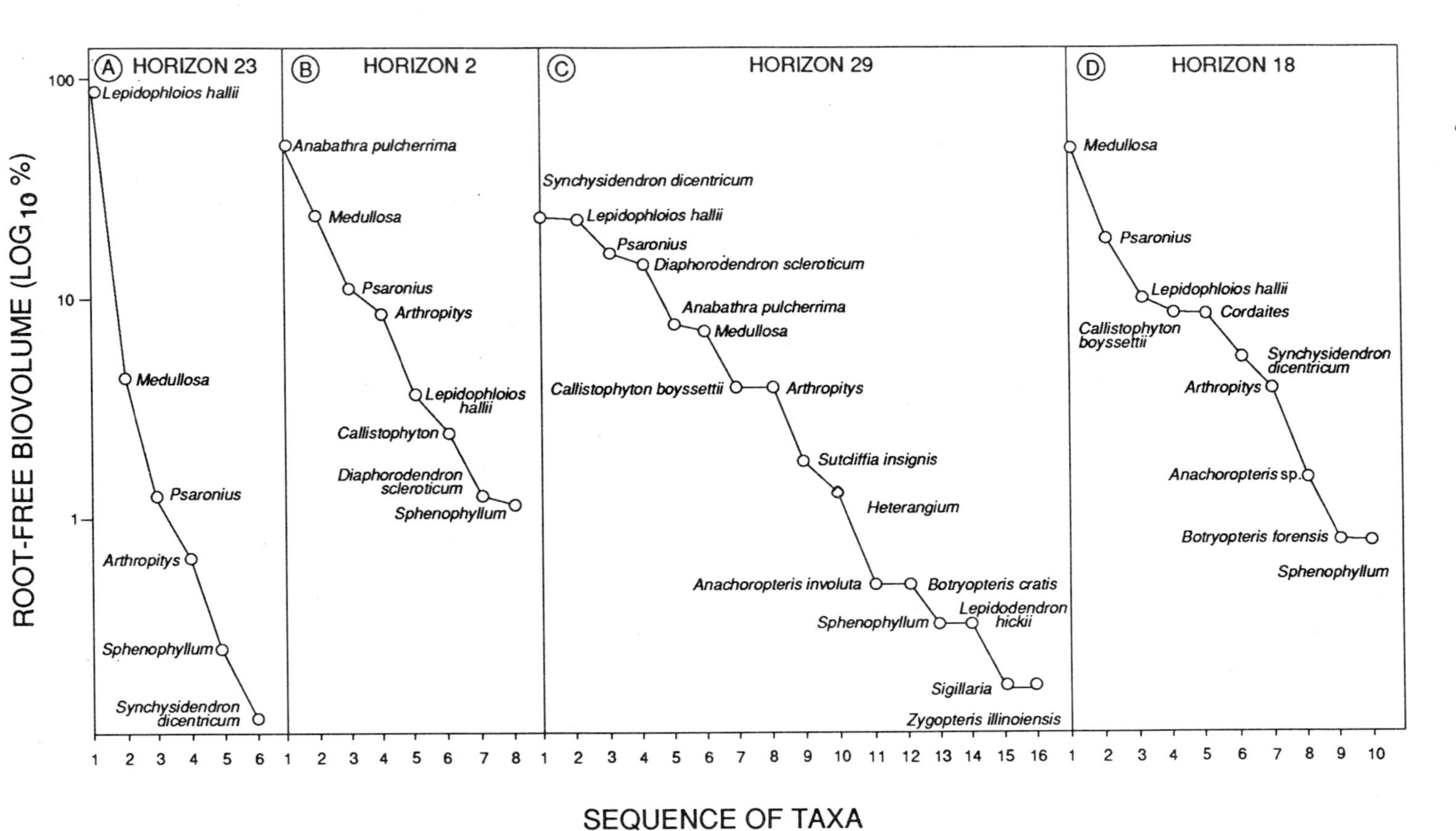

Fig. 2.13 — Dominance diversity curves for root-free biovolume data from four selected, contrasting horizons of the Herrin Coal at Old Ben. Putative whole-plant species are listed from left to right in order of decreasing abundance (Adapted from DiMichele & Phillips 1988, Figs 11–12.)

2.9 CATENARY STUDIES (CARBONIFEROUS, ADPRESSION/MOULD-CAST/ PETRIFACTION)

The aforementioned studies have focused on data gathered from vertical successions (including cores). Such exposures represent one spatial dimension plus the fourth dimension, time. Occasionally, fortuitous exposures (usually the floors or roofs of quarries and mines, and thus dominantly in the Upper Carboniferous) expose an extensive portion of a bedding plane bearing plant remains. These minimize the time spanned by the fossil assemblage while revealing the second lateral spatial dimension. Moreover, some such exposures provide limited information on the third (vertical) spatial dimension, and hence on ecosystem stratification. On the rare occasions when sufficient areal extent of the bedding plane is exposed, a fossil landscape (palaeocatena) is revealed and any fossil plant assemblages present can be analysed using a range of quantitative techniques more familiar to neoecologists (e.g. Digby & Kempton 1987). The types of data obtained are determined primarily by whether the fossils are dominantly autochthonous (*in situ*) or parachthonous (see section 2.2.2).

2.9.1 Autochthonous

Autochthonous assemblages are both more readily identified and more potentially informative than transported plant parts. Roots, rootstocks, rhizomes and upright stem-bases occurring in palaeosols are strong evidence of preservation in growth position (cf. Fritz 1980, Retallack 1981, Fritz & Harrison 1985). However, recent studies based on direct observation have demonstrated that upright stem bases alone are insufficient evidence for unequivocal assertions of autochthony, as they can be transported upright by water, ice or mass flow. Severely truncated large-diameter roots and trunks provide circumstantial evidence of transport (Fritz & Harrison 1985, Fritz 1986), though the absence of a palaeosol is perhaps more telling. Rarely, both soil and *in situ* plants are transported (e.g. as mass flows or rafted peats).

Unfortunately, the individual species comprising autochthonous assemblages are less readily identified. Rooting systems alone can rarely be identified beyond class or ordinal level. Attachment to stem bases greatly improves identification potential, but the diagnostic (unique derived) character states that distinguish whole-plant species are still likely to be absent. Moreover, the generally poor preservation of assemblages means that many of the species originally present in the community, particularly herbs, may be absent from the autochthonous assemblage.

Thus, authochthony *s.s.* refers primarily to rooting organs plus whatever portions of the aerial organs of the plant have persisted (if any). Although there is no theoretical barrier to studying small-bodied plants on palaeocatenas, tree stumps are more readily preserved and more easily identified, and thus have attracted most attention from palaeobotanists.

Examples are the stands of the calcareously petrified stumps of the tree-sized lyginopterid pteridospermalean *Pitus* that occur in the Upper Tournaisian of Weaklaw Rocks, south-east Scotland (Gordon 1935, Scott 1990) and the Mid-Visean of Kingwater, northern England (Long 1979, Scott & Rex 1987); some show growth rings in the wood (Creber & Chaloner 1984). Unfortunately, exposures at both localities are insufficiently extensive to allow quantitative studies; this also applies to the linear coastal sections in the Mid-Upper Visean of Laggan, SW Scotland that expose the petrified *in situ* stigmarian rootstocks of the tree-lycopsid *Lepidophloios wuenschianus* (Walton 1935, Scott 1990).

Nonetheless, it is these stigmarian rootstocks (albeit in their more common mode of preservation: bulk-replaced casts with optional adpressed organic rinds) that have facilitated the most detailed Palaeozoic catenary reconstructions. Despite sporadic reports from the Upper Devonian (e.g. Grierson & Banks 1963) and Lower Carboniferous (e.g. MacGregor & Walton 1972), most examples of *in situ* tree-lycopsid rootstocks have been found in Upper Carboniferous strata (40 occurrences are listed in Table 1 of Gastaldo 1986a; criteria for identifying stumps as *in situ* are given by Gastaldo 1984). The intriguing modes of formation of rootstock casts and their subsequent influence on local sedimentation have attracted considerable attention (Rex & Chaloner 1983, Gastaldo 1986a, b, DiMichele & DeMaris 1987, Wnuk & Pfefferkorn 1987). Rapid decay of axial tissues (especially the nested cylinders of parenchyma) allowed similarly rapid bulk replacement. Most commonly, exposed (and thus reported) lycopsid forests are rooted either in the seatearth immediately below a coal seam (the final community prior to the onset of peat accumulation) or in the uppermost horizons of a coal (the final community to occupy the peat substrate before substantial clastic deposition resumed). Typical clastics in such roof-shale sequences coarsen upwards; thus, the shorter the length of trunk that remains attached to the rootstock, the finer is the infilling sediment and the more probable is preservation of an adpressed organic rind. However, the main palaeoecological interest of 'stump forests' lies not in their mode of preservation but in the information that they provide on the composition, density and residence time of the forest stands.

For the first example, I shall return to the Herrin Coal (Westphalian D) of S Illinois. DiMichele & DeMaris (1987) mapped a rootstock assemblage across the roof of the Orient Mine, 12 km E of the Walshville palaeochannel (Fig. 2.14(a)). Stumps were rooted in the coal but enclosed and replaced by the bay-fill lithofacies of the overlying Energy Shale. Data recorded for each stump included its identity, diameter (a problematic measure owing to disintegration and distortion), length of any attached trunk base, and its deviation from the vertical; nearby, orientations of adpressed fallen trunks were also recorded. Gastaldo (1986a) recorded similar data (though with greater detail on the dimensions of rhizomorph branches and their generally shallow angle of soil penetration) from a stump assemblage at Brooksville, N Alabama. Here, 42 tree-lycopsid rootstock casts occupied an area 240 m by 40 m; they were rooted in the uppermost horizon of an interdistributary-bay shale immediately subjacent to the Upper Cliff No. 2 Coal (Langsettian).

At the Orient locality, all but two of the rootstocks possessed the flared bases characteristic of arboreous lycopsids; they were tentatively identified as true *Lepidodendron* (*L. hickii*), a frequent inhabitant of clastic-rich swamps (they were subsequently identified as *L. jarasewskii* by B. A. Thomas, pers. comm. to W. A. DiMichele). Stump diameter showed a near-normal distribution, with a mode of 100–110 cm and range of 55–144 cm. The recorded density of 33 rootstocks in 130 m^2 (*c*. 1800 per ha) occupied *c*. 10% of the total ground area; this value was extrapolated

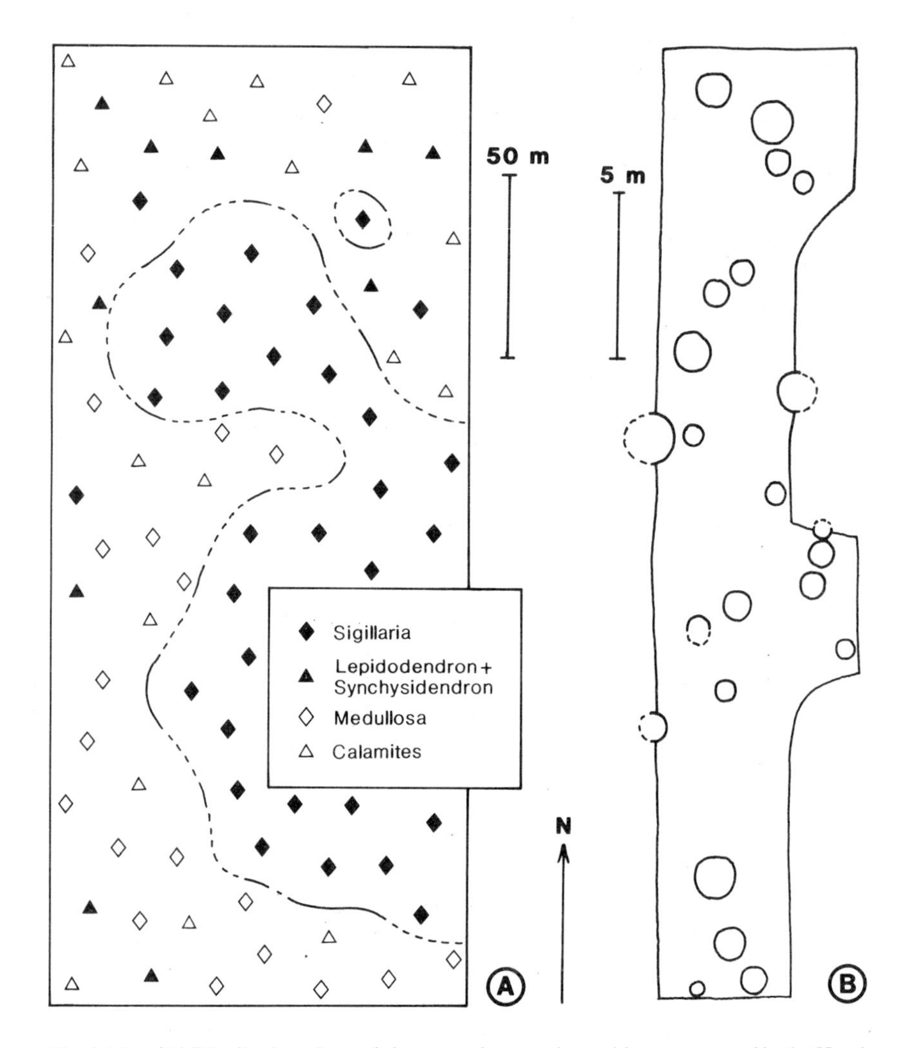

Fig. 2.14 — (A) Distribution of autochthonous arboreous lycopsid stumps rooted in the Herrin Coal, exposed in the roof of a gallery of the Orient Mine. (Redrawn from DiMichele & DeMaris 1987, Fig. 2). (B) Distribution of parachthonous adpression form-genera in the roof-shale of the Springfield Coal at AMAX, showing ecotones. (Adapted from DiMichele & Nelson 1989, Fig. 1.)

to a still substantial *c.* 7% at the breast height level preferred by forest neoecologists (e.g. Burnham *et al.* in press).

DiMichele & DeMaris (1987) used nearest neighbour analysis (R) to test the distribution of the Orient Mine rootstocks relative to extremes of contagion (R=0) and uniformity (R=2.15; for random distributions, R=*c.* 1). In addition, they calculated this statistic for an assemblage of 73 lycopsid stumps from Westphalian strata at Parkfield Colliery, northern England (Beckett 1845), and I have obtained an R value of 0.0 for the 11 stumps of the classic Lower Namurian 'Fossil Grove' of

Victoria Park, SW Scotland (MacGregor & Walton 1972). Gastaldo (1986a) partitioned the elongate exposure at Brooksville into three areas for spatial analysis, sampling 12, 9 and 10 non-vitrainized stumps respectively along a north–south transect. As some stumps were believed to have been eroded, nearest neighbour analysis was rejected in favour of a chi-square test applied to a Poisson distribution. The results suggested slight but statistically significant uniformity at the Orient Mine, randomness at Parkfield, and slight contagion at Victoria Park. At Brooksville, the central plot exhibited significant contagion while distributions in the remaining plots were random.

This preponderance of random and near-random stump distributions led DiMichele & DeMaris (1987) to suggest that the arboreous lycopsids studied by them were rapidly colonizing opportunists that experienced minimal interactions among individuals; indeed, many remain closely juxtaposed (Gastaldo 1986a, Fig. 2, DiMichele & DeMaris 1987, Fig. 2). The Orient Mine stand was envisaged as even-aged, having completed its monocarpic cycle of determinate growth, massive late-stage reproduction, and consequent death. Thus, the low-energy deposition of clastics on the peat did not kill the trees, though it did prevent establishment of their offspring. Long after death, the lycopsid trunks fell in random orientation, buckled and twisted following subaerial degradation. In this scenario, the initial establishment of the stand and its failure to re-establish a further generation reflect changing levels of environmental stress and disturbance, notably flooding of the habitat and associated influxes of clastics. All of the above studies revealed near-monotypic lycopsid stands, suggesting that the stresses were sufficiently persistent and severe to prevent establishment of understorey species.

2.9.2 Parachthonous

The palaeocatena approach has also been applied to parachthonous adpression assemblages rich in large plant organs, notably the fallen trunks (and, in some cases, fronds) of coal-swamp trees.

In the Westphalian D of the Bernice Basin at Lopez, NE Pennsylvania, Wnuk & Pfefferkorn (1987) studied an *in situ* rootstock assemblage comprising several arboreous lycopsid species, together with abundant adpressed litter in the associated rooted grey siltstone and overlying laminated black shale (this in turn grades into the B-Coal). Several 0.25-m^2 quadrat samples taken through 25 cm of these fine-grained sediments revealed a much greater abundance of lycopsid rooting organs, pteridospermalean leaves, and calamite and sphenophyllalean debris in the siltstone, but more abundant lycopsid leaves in the shale. Both assemblages were interpreted as litter layers, the former accreted on the landscape occupied by the forest community, the latter formed after the deposition of the shale during the long-term flooding that extinguished the vegetation and eventually led to peat accumulation. Adpressed pteridospermalean trunks within the black shale show weaker preferred orientation than co-occurring tree-lycopsids (Fig. 2.15); comparison with orientations of fallen trees and sedimentary sequences in modern analogues suggests that the lycopsids were felled post-mortem and *en masse* by high velocity winds, crushing the lower-growing pteridospermaleans as they fell (cf. Gastaldo 1990).

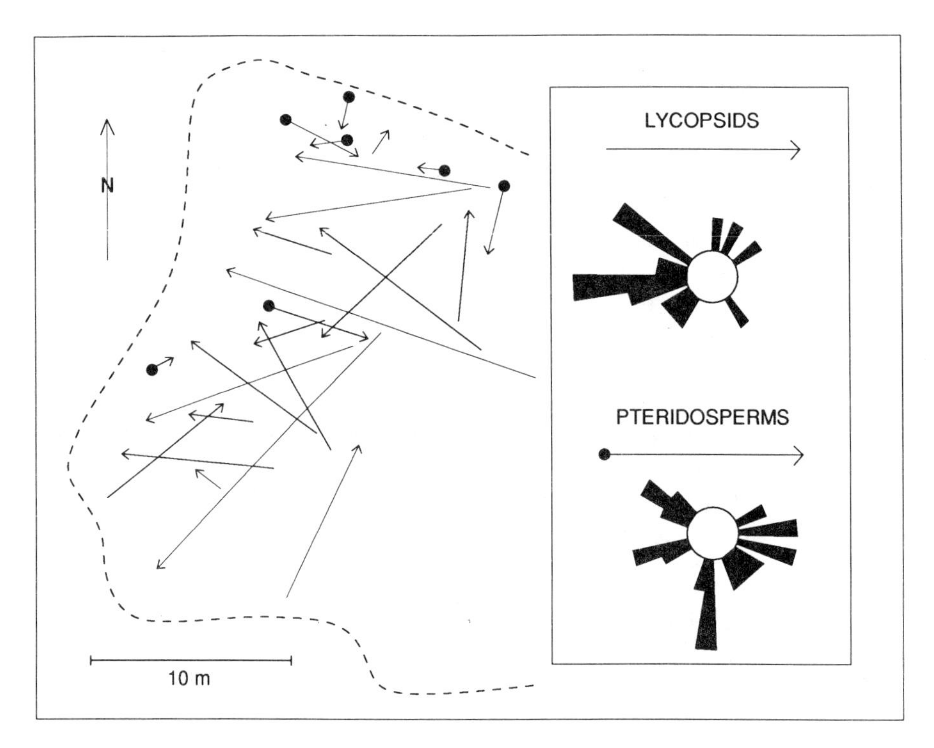

Fig, 2.15 — Orientations of adpressed arboreous lycopsids and medullosan pteridosperms in the B-Coal underclay of an opencast mine near Lopez, Pennsylvania. Only the larger pteridosperm axes are shown to avoid confusion; a larger number contributed to the rose diagram. (Adapted from Wnuk & Pfefferkorn 1987, Fig. 3.)

Bernice Basin bedding-plane assemblages also allowed reconstruction of the gross morphology of arboreous lycopsids of the genera *Diaphorodendron*, *Synchysidendron* (Wnuk 1985) and *Bothrodendron* (Wnuk 1989), and several medullosan pteridospermaleans (Wnuk & Pfefferkorn 1984). DiMichele & Nelson (1989) relied on comparison with these reconstructions and others based on anatomically preserved coal-ball species (Table 2.3) when interpreting a 2.5-ha bedding-plane assemblage in the roof-shale immediately overlying the upper Westphalian D Springfield Coal of the AMAX Mine in south Illinois. As in the Herrin Coal study of DiMichele & DeMaris (1987), the community occurred close to a palaeochannel (in this case, 0.6 km south-east of the Galatia palaeochannel). Some fossil plants remain rooted in the uppermost horizons of the coal-forming peat but are enclosed by the overlying roof-shale (the overbank/splay Dykersburg Shale); disarticulated axes within the shale appear randomly oriented. The community was regarded as the final stand of the coal-swamp forest, drowned by flooding from the palaeochannel.

Of the 2.5 ha mapped by DiMichele & Nelson (1989), a core area of 0.7 ha yielded a near-monotypic stand of *Sigillaria mamillaris* stumps up to 1 m in diameter, delineated by a sinuous ecotone (Fig. 2.14(b)). Along its south margin, the ecotone was narrow (*c.* 2 m); the *Sigillaria* stand rapidly passed into scrub composed of pteridospermaleans (represented mostly as large fronds of at least one species of the putative medullosan *Neuropteris*) and calamites (pith casts up to 20 cm in diameter of *Calamites* cf. *suckowii*). Scattered, poorly preserved tree-lycopsids were present but did not occur within 30 m of the ecotone. However, the ecotone was much broader to the west, north and north-east, where there were fewer pteridospermaleans and more tree-lycopsids; *Lepidodendron* and *Synchysidendron* were admixed with, but showed a negative correlation to, *Sigillaria*. Once again, there was little evidence of ground cover. DiMichele & Nelson (1989) were unable to relate the ecotone to lithological variation, and suggested that differences in microhabitat preference (perhaps related to the depth and/or duration of standing water) caused the observed small-scale area patchiness in the swamp-forest community.

In a larger-scale study, Gastaldo (1985b, 1987) studied several bedding-plane assemblages in a Langsettian deltaic sequence at Clarence, north Alabama. Assemblages from four localities were quantified for both *in situ* tree-lycopsid stumps (n=2–31) and associated adpression assemblages assumed to represent forest floor litter (n=77–258). The localities occurred along a 2.3-km-long, north-west–north-east-oriented transect running parallel to, and immediately north-east of, en échelon clastics indicating the passage of a sizable palaeochannel. The assemblages were interpreted in terms of proximity to the palaeochannel margin (150–800 m) and to a substantial crevasse-splay that entombed a levee-derived assemblage midway along the transect. Comparison of *in situ* stumps (again typically 1 m in average diameter) from the four localities suggest that the swamp supported a near-monotypic stand of *Lepidophloios laricinus*. Close to the crevasse-splay, this was joined by *Lepidodendron aculeatum*, while the levees were dominated by *Sigillaria* cf. *elegans*. Putative fossil litter samples indicate that appreciable quantities of calamites and pteridospermaleans (dominantly two form-species of the putative medullosan shrub *Neuropteris*, but also the putative lyginopterid liana *Lyginopteris hoeninghausii*) occurred on the better-drained levee and crevasse-splay substrates (Gastaldo 1987). These observations reinforce the now traditional perception of the spatial relationships of major plant taxa in coal-swamps: some arboreous lycopsid genera in the swamps, others on comparative topographical 'highs' such as levees with pteridospermaleans and calamites, which also occurred along channel margins.

2.10 MANY PROBLEMS, SOME SOLUTIONS

This final section builds upon the previous sections through a series of vignettes that address key methodological and interpretational topics. Some of the issues discussed have been considered by many other authors, others have been consistently overlooked; likewise, some of the opinions expressed are mainstream to modern plant palaeoecology, others are highly idiosyncratic. All are incompletely developed; they are included to prompt further debate rather than masquerade as definitive statements.

2.10.1 Completeness of the plant fossil record

The most serious interpretative constraint on Palaeozoic palaeoecology is the patchy record of biotas in the terrestrial realm, which experiences net erosion (this contrasts with the more complete fossil record in the marine realm, which experiences net deposition). There are four main aspects to actual completeness for any specific time period: the number of fossil plant assemblages present, their distribution through time and space (i.e. degree of concentration within that time period), their average quality and range of preservation, and the range of habitats that they represent. The third and fourth criteria are strongly dependent on the degree of transport typical of the assemblages; as a general rule, mixed assemblages containing both autochthonous and parachthonous materials are most biologically informative, while allochthonous assemblages cannot be apportioned with confidence to specific growth habitats. Quality of preservation, a function of degree of articulation and mode of preservation, determines how precisely fossils can be identified taxonomically (some key Silurian fossils cannot be attributed with confidence to the plant or animal kingdoms!), and how successfully they can be conceptually reconstructed into biologically meaningful whole-plant species.

Fossil plant assemblages are especially sparse in the pre-Devonian, Middle Devonian, and Lower Permian. These are also the intervals of poorest average quality of preservation; most assemblages are allochthonous and few contain petrifactions. The Upper Carboniferous provides the most complete record, as well as exhibiting the widest range of preservational modes and greatest proportions of autochthonous and parachthonous assemblages. Consequently, it has yielded almost all of the credible reconstructions and autecological interpretations of Palaeozoic plants. Similarly, the number of published palaeoecological studies for a period is roughly proportional to the number and quality of the assemblages; there are few for the Upper Silurian–Lower Devonian and Lower Permian but many for the Upper Carboniferous. Typical environments of preservation of plant assemblages differ greatly among geological periods in Laurasia. The Lower Palaeozoic is characterized by fossil preservation in marine strata, and the Devonian by deltaic and fluvial facies. Vegetation from volcanigenic landscapes is usually well represented in the Lower Carboniferous, while the Upper Carboniferous is epitomized by the remains of coal-swamps. It is more difficult to categorize typologically Permian assemblages, though seasonal fluvial systems figure prominently. Poor representation of some physiographical regions and habitats leaves important questions unanswered, such as the times when salt-marshes (Gensel 1986) and uplands *sensu stricto* (Remy 1975, Mapes & Gastaldo 1986) were first colonized by land plants.

In addition, available data are determined by the perceived rather than the actual completeness of the fossil record; that is, by the intensity of study of fossil plant assemblages of a particular time period in a particular physiographical region. This factor more than any other prompted the strongly European/American bias of this chapter.

2.10.2 Reconstructed whole plants as pivotal entities

The paleobiological value of conceptual whole plants is immense (Chaloner 1986, Galtier 1986, Bateman & Rothwell 1990). Their significance stems primarily from the fact that all their component organs can be characterized to some degree. Petrified whole plants generally provide more characters than adpressions, though the information recovered from the latter can often be increased through macroscopic studies of morphology via morphometrics (e.g. Scheihing & Pfefferkorn 1980) and microscopic studies of macerated cuticle (e.g. Cleal & Zodrow 1989) and spores (e.g. Fanning *et al.* 1988).

The wide range of characters renders whole plants the most appropriate focus for palaeobotanical taxonomy (cf. Thomas & Brack-Hanes 1984, Meyen 1987, DiMichele & Bateman in press). They are used (albeit more often implicitly than explicitly) as core species: templates that provide the foundations for higher taxa and the comparative basis for identifying less articulated and/or less well preserved partial plants (satellite taxa). They improve the probability of correctly synonymizing undesirable form-species based on ontogenetic or ecophenotypic variants, which artificially inflate floristic lists. Key lower taxa initially labelled as intermediate between higher taxa (e.g. the 'pre-fern' *Rhacophyton*: Matten 1974, Scheckler 1986b) often appear less so as data accrue. Core species are essential basic units for phylogenetic reconstruction and thereby for testing hypotheses of modes of evolution. Most importantly from a paleoecological perspective, whole-plants are necessary prerequisites for confident autecological interpretations. When integrated, these allow synecological analysis of communities, which in turn leads to stronger palaeogeographical and palaeoclimatological inferences (Fig. 2.1).

A wide range of evidence can be used to aid whole-plant reconstruction (e.g. Bateman & Rothwell 1990). Overall, the probability of success is greatest when assemblages (1) occur repeatedly as a series of closely spaced horizons containing the remains of a single community, (2) have experienced minimal disarticulation, (3) have experienced minimal transport, (4) are exceptionally well preserved, and (5) are of low species diversity. It is also aided when species present are (6) only distantly related to each other (and thus morphologically distinct), and (7) primitive (and thus composed of relatively few organs).

Criteria (2), (3) and (4) are met by few assemblages. Criteria (6) and (7) are mutually antagonistic, though both favour analysis of early land-plant communities rather than their later descendants. The species-poor assemblages required by criterion (5) generally yield less profound synecological interpretations than do species-rich communities. Criterion (1) similarly reveals a potential conflict between assemblages optimal for whole-plant reconstruction and those optimal for synecological interpretation; ideally, successive horizons represent the same community in the former but a progressive series of communities in the latter. However, although the goals of palaeoecological interpretation and whole-plant reconstruction are not identical, they are sufficiently close to be of great mutual benefit.

2.10.3 Application of uniformitarian principles to Palaeozoic floras

Uniformitarian extrapolation from the present to the past, from dynamic observation to static inference, has played two distinct roles in Palaeozoic palaeobotany: (1) morphotypic comparison of living and fossil plants to interpret functional morphology, and (2) assumptions of constancy (or at least of fluctuations that lack long-term directionality) in environmental parameters. When combined, these two assumptions generate a third, synthetic package of uniformitarian concepts: taphonomic reconstruction.

Much ecological interpretation of post-Palaeozoic floras, whether megafossil or microfossil assemblages, depends upon morphological comparison with extant species whose behaviour can be observed directly. The more distantly related the extant and extinct species under comparison, the less reliable is the method; consequently, its applicability to Palaeozoic floras, which differ from modern communities at the taxonomic levels of class and order, is minimal. This taxonomic extrapolation is especially problematic for miospore studies, where few characters are available; not only is comparison with extant species invalidated, but also relatively few Palaeozoic spore-species have been traced to source whole-plant species (e.g. Courvoisier & Phillips 1975, Willard 1989).

An increasingly popular alternative to phylogenetic extrapolation is to ignore evolutionary relationships and instead use functional morphology as a vehicle for comparing the ecological roles of phenotypes (ecomorphotyping *sensu* Wing *et al.* in press) and thereby identifying ecological analogues ('vicars'). Species replacement means that ecomorphotypes are likely to persist longer than species *per se* (DiMichele *et al.* in press); the ecological role outlives the taxonomic player. However, even taxon-free ecomorphotypes cannot be compared in a vacuum; comparison of communities through time requires that they should be drawn from similar environments of growth (habitats). This is problematic, given that most Palaeozoic plant fossil assemblages (especially outside the Upper Carboniferous) are allochthonous and thus consigned to the realm of taphonomic inference (Spicer & Greer 1986, Gastaldo 1988, Burnham 1989, Burnham *et al.* in press, Behrensmeyer *et al.* in press).

Hutchinson's (1965) seminal paper on ecological niches appeared in a volume cleverly titled *The ecological theatre and the evolutionary play*. This attractive analogy has been over-extended in some studies, to the point where evolution is perceived as a dynamic process proceeding on a static stage (even on a stage, props and lighting change with every scene). Other studies consider environmental changes but view them only in the context of the present environment. Not only does this overlook the potential for rare but profound events beyond our observable time-frame (a topical example is the much-vaunted extraterrestrial bolide impact), it also ignores progressive, long-term changes in environmental parameters such as atmospheric oxygen and carbon dioxide concentrations (e.g. Robinson 1989, Berner 1990), which in turn influence the frequency and magnitude of other ecologically important phenomena such as wildfire (e.g. Cope & Chaloner 1985). It similarly overlooks modifications of the physical environment by the plants themselves, such as soil development and stabilization (Wing 1984, Retallack 1985). Environmental uniformitarianism, like taxonomic and functional uniformitarianism, has its limits, and it is important that these should constrain palaeoecological interpretations.

2.10.4 Uniformitarianism and phylogenetic reconstruction

As noted in the previous section, functional morphology is the cornerstone of morphotypic comparison. However, function in fossil plants can only be inferred, and such inferences must assume adaptive value. I would argue that the further back land-plant lineages are traced, the smaller is the average proportion of their morphological characters that are likely to have been truly adaptive. The early history of land plants is rife with apparently 'experimental' morphological radiations in certain features; a good example is the remarkable range of esoteric microphyll leaf morphologies exhibited by Middle Devonian lycopsids (Gensel 1986, Hueber in press). Moreover, many other features probably evolved adaptively for one function but later were co-opted for other functions; one such example is plant cuticle (Knoll *et al.* 1986). Another more behavioural example is the apparent origination in plants of defences against herbivores long before herbivory had become a serious problem (DiMichele *et al.* in press). Only one (more often the later) function is likely to be detected (see Donoghue 1989).

Rather than ignore evolutionary relationships, an increasing number of functional morphologists attempt to subtract historical constraints from assessments of adaptation (e.g. Pagel & Harvey 1988). As this can only be satisfactorily achieved by 'knowing' the evolutionary relationships of the species in question, access to a viable phylogeny of those species becomes crucial. Similarly, the apparent iterative evolution of features of great ecological utility (e.g. branches, leaves, wood, heterospory, the seed habit *s.l.*) can only be tested via phylogenies; once identified as apparently iterative, we can search more carefully for detailed differences among superficially similar features that might support hypotheses of their independent origins. Given the sparseness of the Palaeozoic plant fossil record in general, and of reconstructed whole plants in particular, methods of phylogeny reconstruction that give priority to stratigraphical relationships over morphological evidence (e.g. stratophenetics: Gingerich 1979) have little applicability. Morphologically based cladograms (e.g. Wiley 1981, Brooks & McClennan 1991, Wiley *et al.* in press) are preferred; they are constructed without reference to the stratigraphical record, which thus remains an independent test of the phylogenetic hypothesis.

Some particularly interesting hypotheses can only be tested in the presence of a rigorous phylogeny of land plants, an equally rigorous classification based on that phylogeny, and knowledge of the taxonomic composition of communities. If the history of land plants is viewed as a series of taxonomic radiations, the radiations decrease in what I shall term 'taxonomic magnitude,' occurring at the level of division in the Silurian, class in the Devonian, and order and family in the Carboniferous and Permian. That is not to say that they decrease in magnitude at the species level; quite the reverse. Thus, the overall pattern of radiations is essentially fractal (Bateman in prep.).

This may explain the low depths of guilds (ecomorphically similar species coexisting in a community) in early land-plant communities noted by DiMichele *et al.* (in press). Diversity in gross morphology and architecture (as reflected in diversity of higher taxa) increased more rapidly than diversity in ecologically interactive units (species and lesser demographic entities); thus, early niches were relatively broad but there were relatively few potential occupants. It would also explain 'taxonomic overprinting' of ecomorphically delimited niches perceived by Niklas (1986) and DiMichele *et al.* (in press), who suggested that niches (and even communities) were partitioned much more along taxonomic lines in the Palaeozoic than they are today. This argument is attractive, though the magnitude of the difference may have been exaggerated.

2.10.5 Relationship between taxonomic and ecological hierarchies

Few plant palaeoecological studies explicitly consider the hierarchical level at which ecological interactions take place. Biotic interactions, whether with other biota or with the environment, inevitably occur at the level of individual organisms. Beyond this, behavioural cohesion is expected among conspecific individuals within a local population, which are the primary units of microevolution and thus of adaptation (and the primary study units of neoecologists). In contrast, the ecological tolerances of any entity higher in the demographic taxonomic hierarchy (*sensu* Bateman & Denholm 1989) are little more than the sum of the tolerances of its component populations; in other words, supraspecific taxa do not possess niches. Although a palaeoecological statement based on study of a population can be extrapolated through higher taxonomic levels, this results in an increasing risk of either generalizing the statement to the extent that it becomes meaningless or omitting (either overtly or covertly) taxa that contravene the statement.

Discussions of plant palaeoecology rarely focus on subgeneric levels and often generalize at the class/order level (a good example in this chapter is the final sentence of section 2.9). Hence, a potentially valuable observation such as '*Calamites suckowii* dominated point-bars of the upper Westphalian D Galatia palaeochannel' is often translated into a much less meaningful statement such as 'sphenopsids dominated Carboniferous point-bars'. Moreover, the more generalized statements tend to become self-fulfilling prophesies. Once it has become generally accepted that sphenopsids occupied point-bars, this assertion can be perpetuated in the face of strong contradictory evidence, particularly if it can be supported using uniformitarian arguments based on evidence from extant communities. This too can be misleading; for example, most of the species of *Equisetum* native to Britain prefer riparian habitats (including point-bars!), but the greatest biomass of *Equisetum* probably occurs in cornfields, where *E. arvense* is a pernicious weed! Even if within-clade ecological replacements were more common in the Palaeozoic than they are today (DiMichele *et al.* in press), the environment(s) of growth represented by each fossil assemblage should nonetheless be assessed on its own merits, preferably using a range of biotic and abiotic evidence, rather than routinely subordinated to preconceived generalizations.

2.10.6 Evolutionary patterns through time

The diversity pattern within a clade through time is a function of species origination and extinction; dominance of the former causes radiation, dominance of the latter causes contraction. Given the sporadic record, there is little chance of detecting a major clade immediately after its origination; it must first extend its geographical and ecological range to the point where it is most likely to be sampled. This is most readily achieved by taxonomic radiation. Assuming constant probability of preservation through time (obviously a rash assumption), it is the acme of a clade that will yield the most unambiguous data (for example, the best whole-plant reconstructions). Evidence diminishes as a clade is traced backward in time; the ecological role and taxonomic affinities of the fossil plant fragment become more equivocal (often, this also applies to tracing a clade forward in time, particularly if the clade is now extinct). A good example of this problem is the ambiguous evidence for the origin of the true tracheophytes (Gray 1985, Edwards & Fanning 1985, Kenrick & Crane in press). Another is the possible persistence of the progymnospermopsids, which appear to have dominated many communities in the Late Devonian and into the Early Carboniferous. The possible ecological vacuum in the earliest Carboniferous communities left by the end-Devonian extinction of the large-bodied tree *Archaeopteris* has attracted much attention (Beck 1964, Scheckler, 1986a, DiMichele *et al.* in press), but *did* the archaeopteridaleans become extinct at that time? Common Lower Carboniferous foliar form-genera such as *Rhacopteris* could easily represent progymnospermopsids rather than pteridospermaleans. On the other hand, the best example of putative progymnospermous reproductive material from the Lower Carboniferous (Walton 1957) is not wholly convincing. The problem is largely one of identification; a progymnospermopsid is little more than a primitive seed plant that lacks seeds. In disarticulated plant assemblages rich in early pteridospermaleans, it is even more difficult to prove that a particular plant axis did *not* bear seeds than to prove that it did (Bateman & Rothwell 1990).

Given these difficulties, much attention is paid to temporally outlying occurrences of particular clades. Some greatly extend the recorded range of a clade forward in time; for example, that of the sphenopsid family *Archaeocalamitaceae* was recently prolonged by 55 Ma (Mamay & Bateman 1991). Other outliers push back a previously widely accepted earliest occurrence. The classic example here is the Lower *Baragwanathia* assemblage of south-east Australia. Although relatively diverse and of typical Siegenian–Emsian composition, this has been dated on dubious graptolite evidence (Garratt & Rickards 1984) as Ludlow, whereas all other assemblages of this age contain only non-vascular plants and cooksonioids (Fig. 2.5). In this case, it is more parsimonious to reject the suspiciously early date given to the fossils than to radically alter our account of the pattern and timing of land-plant evolution (cf. Knoll *et al.* 1986, Hueber in press). Our perception of the timing of major radiations within a clade, and thus its rise to ecological importance, is less prone to modification by retrogressive range extensions than is our perception of the timing of the origination of that clade. However, retrogressive extensions *do* influence our perception of the time-lag between the origination of a clade and its rise to dominance; this in turn affects our perception of *how* dominance is achieved. Current evidence suggests that such lags often span tens of millions of years; these relatively long periods are consistent with the hypothesis that radiation cannot be achieved by competitive replacement alone but requires severe environmental perturbations to free niches for colonization (Knoll 1984, Scheckler 1986a, DiMichele *et al.* 1987, in press).

2.10.7 Diversity within plant communities through time

Recent descriptions of the overall diversity of land plants through time (Knoll *et al.* 1979, Niklas *et al.* 1980, 1983, 1985; see also Signor 1990) led logically to comparable quantitative asessments of community diversity (Niklas *et al.* 1985, Knoll 1986, Niklas 1986). To examine the pros and cons of such procedures, I shall focus on Knoll's (1986) curve.

Data were extracted from the published literature on adpression assemblages at the scale of 'floral lists', each typically reflecting a sample from an extensive geographical area and strata of formation or member thickness (Table 2.4). Mean number of species per 'flora' was then taken as 'an imprecise proxy for community diversity' (Knoll 1986: 132). The resulting curve shows three successive plateaux, respectively of approximately four, 11 and 26 species per 'community' (Fig. 2.16).

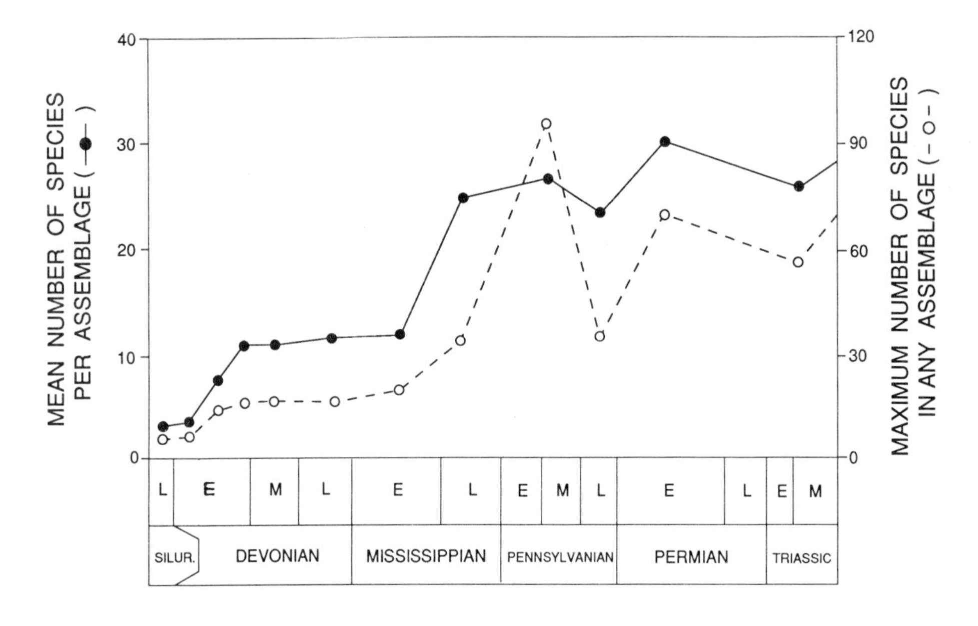

Fig. 2.16 — Mean (solid) and maximum (dashed) species-level diversity curves for adpression-bearing localities through the Palaeozoic. (Adapted from Knoll 1986, Table 7.1, Fig. 7.1.)

The plateaux are separated by sharp increases in diversity through the Lower Devonian (reflecting the radiation of the division Tracheophyta) and across the mid-Mississippian boundary (reflecting the radiation of the Pteridospermales, the earliest order of the class Gymnospermopsida); there are no sharp decreases. Maximum community diversity tracks mean diversity closely at 150–200% of the mean, except for a rise to 350% in the well-studied (or perhaps taxonomically over-split: Knoll *et al.* 1979) mid-Pennsylvanian coal-swamp assemblages.

Table 2.4 focuses on species-level community diversity in the interval with which I am most familiar, namely the Lower Carboniferous (lower Mississippian). The new data contrast with those of Knoll (1986) in that the geographical area sampled is smaller (north-west Europe rather than global equatorial) and the assemblages studied are petrifactions rather than adpressions; the former yield more diagnostic characters and thereby a more reliable taxonomy. Methods used in the two studies to reduce form-species lists to putative whole-plant species were broadly similar but differed in detail. Both used the mis-named 'minimum species concept', which reflects the largest number of form-species in any category of organ from the locality in question. Presumably, miospores were excluded by Knoll from the range of form-

Table 2.4— Comparison of assessments of species-level diversity in Lower Carboniferous plant assemblages. Data for the present study were abstracted from a database compiled by the author for the Smithsonian Institution's 'Evolution of Terrestrial Ecosystems Consortium'; assemblage lists known to be very incomplete (either because of poor preservation or superficial study) were omitted. Note that the present data and those of Knoll (1986) are not wholly comparable (see section 2.10.7). For Oxroad Bay see also Table 2.2.

Nature of fossil plant assemblages				Diversity (no. of species)				
				Oxroad Bay assemblage	All sampled assmblages			
Species concept	Diameter of area sampled	Stratigraphic unit	Mode of preservation		Mean	S.D.	Max.	*n*
Knoll (1986, Table 7.1): Tropical/subtropical, ?global								
Whole plant: Broad	>1 km	Formation/ Member	Adpression	—	11.9	3.5	19	15
Present study: Tropical, north-west Europe only								
Organ	<1 km	Member	Petrifaction: All modes	43	16.1	9.2	43	21
Whole-plant: Restricted	<1 km	Member	Petrifaction: All modes	12	7.9	3.4	16	21
Whole-plant: Restricted	$\leqslant$1 km	Bed	Petrifaction: All modes	8	6.5	2.5	11	15
Whole-plant: Restricted	$\leqslant$1 km	Horizon	Petrifaction: All modes	5	cannot be assessed from available literature			
Whole-plant: Restricted	$\leqslant$1 km	Horizon	Petrifaction: Restricted modes	4				

species categories under consideration; in the present study, megaspores (including ovules) and isolated sporangia were also excluded from 'minimum species' calculations as being too prone to long-distance transport (for example, note the superfluity of pteridospermalean ovule-species relative to whole-plant species in units 2 and 4a at Oxroad Bay: Table 2.2). In addition, current knowledge was used to synonymize superfluous form-species. The more rigorous petrifaction taxonomy and more narrowly defined 'minimum species' together are probably sufficient to account for the differences between the two studies in mean and maximum whole-plant species numbers per locality: respectively 12 and 19 in Knoll (1986) versus 8 and 16 in the present study. However, beyond this broad agreement lies another important question: is this the most appropriate spatio-temporal scale for determining community diversity?

Figures for species diversity at a single locality, Oxroad Bay (Table 2.4), demonstrate that the scale of sampling and interpretative framework are crucial. The locality covers an area of *c.* 6 ha and spans *c.* 40 m of intermittently plant-bearing strata, and assemblages appear dominantly parachthonous (Bateman & Scott 1990). It is the most diverse of all Lower Carboniferous petrifaction-bearing localities *at the form-species level*, but the 43 recorded form-species merely reflect unusually intense study of the locality. Application of the restricted miminum species concept indicates that these represent only 12 whole-plant species (Bateman & Rothwell 1990), a figure higher than the mean for all localities but lower than the maximum at this taxonomic level. However, reducing the spatiotemporal level of study to a single bed (broadly a single landscape) reduces the maximum number of putative whole-plant species to eight, close to the mean value for all localities at this level of resolution (6.4: Table 2.4). Further reduction of stratigraphical scale to a horizon (broadly, a single depositional event) further reduces this figure to five; moreover, within a single horizon, no more than four species show the same mode of preservation.

I regard this last, most narrowly delimited figure as the closest approximation to the maximum number of species at Oxroad Bay likely to have occupied a single habitat at a single moment in time. Thus, I contentiously define a community as a group of populations co-existing within a narrow and definable set of environmental conditions, and consider this to be the most palaeoecologically meaningful scale of analysis. Because data are rarely collected and even more rarely presented at this high level of spatiotemporal resolution, such detail cannot be obtained from the literature; instead, it must be generated *de novo* by researchers intimately familiar with the locality under scrutiny. Moreover, form-species lists are difficult to translate satisfactorily into estimates of whole-plant diversity, even by specialists in assemblages of that particular age and mode of preservation.

2.10.8 Value of and limits to quantification

Several of the above case studies demonstrate the great benefits of quantifying fossil plant assemblages, irrespective of organ type(s) (e.g. megafossil versus microfossil) and preservational mode(s) (e.g. petrifaction versus adpression). Admittedly, more frequent publication of raw data would have enabled reinterpretation using more sophisticated statistical techniques (Kovach 1988) within an improved conceptual framework (Spicer 1988); good data generally outlive good interpretations.

Even good data are no panacea. Perhaps the best measure of ecological success at a particular moment in time is standing biomass. The closest palaeobotanical approximation, coal-ball petrifactions, are not standing but detrital (albeit dominantly parachthonous) and not biomass but biovolume; comparative studies with adpression assemblages from coals suggest that there can be up to an order of magnitude difference in relative compressibility and thus in approximate biomass between different organs of identical biovolumes (e.g. Winston 1988), depending on the volumetric ratio of solid cellular materials versus protoplasm and/or air. Adpression assemblages are essentially two-dimensional and tend to be allochthonous; in such cases, quantitative data reflect only bioareas of hydrodynamically equivalent plant particles (though this can nonetheless approximate relative biomass: Burnham *et al.* in press). Miospore assemblages are even further removed from standing biomass. They are by definition transported, and the proportion of biomass devoted to spore production differs radically among species; indeed, it can differ radically among conspecific individuals, and within a single individual through its life-span (e.g. Phillips & DiMichele in press). In this context, a valuable cautionary note is provided by Bartram's (1987b) combined study of lycopsid megaspore and microspore assemblages in a Westphalian coal; *Paurodendron* and *Chaloneria* achieved maximum frequencies of 95% and 60% respectively in the megaspore assemblages but only 7% and 3% respectively in the corresponding microspore assemblages. Did these genera dominate the communities in question as the megaspores suggest, or were they minor components as the microspores suggest? Were non-lycopsids, incapable of generating dispersed megaspores, predominant? Once acquired, other large databases will undoubtedly reveal similar contradictions between different lines of evidence, and many will prove difficult to resolve satisfactorily. Despite these cautionary notes, contradictions cannot be recognized, let alone addressed, without quantification.

Although quantification of Palaeozoic assemblages now has an established history, quantification of their abiotic contexts is virtually unexploited. A great deal can be learned about environments of growth and/or deposition by applying quantitative sedimentological techniques to determine particle size distributions, mineralogical and geochemical composition, and clast orientation (Bateman & Scott 1990). Together with taphonomy, sedimentology should play a much greater role in future plant palaeoecological studies.

2.10.9 Correcting for habitat

Many neoecological studies rely on controlled experiments involving living organisms where variation in most environmental and biotic factors is minimized, thus greatly simplifying (arguably over-simplifying) the observed interactions. The experimental approach is beyond the reach of the palaeoecologist, as is altering the species composition of the study community. However, environmental variation can still be reduced, simply by focusing on specific growth habitats. Partitioning palaeoecological problems into manageable units improves the precision of the data and thereby the reliability of the interpretations. In particular, vegetational changes that appear gradual on a global scale often appear much more rapid (and more explicable) when studied in a limited range of habitats and/or a restricted geographical area. A good example is provided by the rise of xeromorphic seed-plants through the Late Carboniferous and Permian, which apparently achieved dominance diachronously, both among phytochoria (Euramеria>Angara>Cathaysia: Meyen 1982, Knoll 1984) and among major habitat types within a phytochorion (DiMichele *et al.* 1985, 1987, in press, Scheckler 1986a, DiMichele & Aronson in press). This transition is particularly well documented in wetland assemblages, where it appears to have been driven primarily by diachronous decrease in precipitation and increase in seasonality (Phillips & Peppers 1984, Ziegler 1990).

Wetlands constitute the habitat (or more accurately set of habitats) that, owing primarily to taphonomic factors, represents the best overall compromise between quantity and quality of palaeobotanical information through the Palaeozoic. They are extreme environments: plants are rooted in soils that are exceptionally rich in free water but poor in oxygen, alkalis and most nutrients (Grime 1979, DiMichele *et al.* 1987). Consequently, they support plant species that are highly specialized and have a (perhaps exaggerated) reputation for being evolutionarily conservative. As most swamps are essentially isolated ecological islands (DiMichele & DeMaris 1987, DiMichele *et al.* 1987, DiMichele & Nelson 1989), even gradual environmental changes can extirpate species (or entire communities if sufficiently profound); there is little opportunity to escape by migration. However, the Late Carboniferous wetlands were unusual in being both extensive and persistent. This allowed the evolution of structurally and taxonomically complex communities characterized by finely divided niches (DiMichele *et al.* 1987, in press). Taxonomic composition typically changed by ecomorphic replacement: following disturbance, a particular species was usually succeeded by a closely related species that apparently evolved outside the swamp (DiMichele *et al.* 1985, 1987, Scheckler 1986a). Their persistence also allowed the coal-swamps to act as refugia for clades that were eliminated in more mesic extra-swamp habitats; this hypothesis has been extended to encompass earlier swamps, such as preserved in the Lower Devonian Rhynie Chert (Knoll 1985a).

Although these specific assertions on swamp palaeoecology require further testing, their broader implication is clear; the dynamics and persistence of a community should be assessed within the constraints of a specific habitat. Moreover, the habitat itself should be carefully defined, as has been achieved for wetland habitats by Moore (1987).

2.10.10 Correcting for time and scale

Radiometric dating accesses geological rather than ecological time. Consequently, within the ecological time-frame (typically less, and usually much less, than a million years), the palaeoecologist is constrained to estimates of relative time rather than absolute time (Grant 1986). It has become popular to perceive moments in relative time as three-dimensional time-planes (in vertical section, these are viewed as time-lines). The closest stratigraphical approximation to a time-plane is a fossil landscape (Palaeozoic studies of such palaeocatenas are reviewed in section 2.9), though in practice a landscape can represent a considerable period of time, given the infrequency of deposition in most terrestrial environments. Also, a landscape is three-dimensional rather than planar, at least to the depth penetrated by bioturbation and soil formation. Tracing a particular landscape laterally increases the probability of finding autochthonous (or preferably mixed autochthonous–parachthonous) plant assemblages, rather than having to guess environments of growth from information gathered in the depositional environments of allochthonous assemblages. It also reveals lateral variation in vegetation and habitats, thereby allowing delimitation of communities in vegetational mosaics (Bateman & Scott 1990). Within a specific habitat, lateral study can avoid atypical sampling, for example, of a single fallen tree, whether senescent or lightning-struck, in coal balls (Winston 1988).

However, understanding of phenomena such as succession (vegetational changes through time in the same space; that is, in the same habitat) requires evidence of repetition of ecological patterns through time; sequential plant-bearing landscapes are needed to identify such successional recapitulation. Even then, seral succession can be difficult to distinguish from longer-term quasisuccession (*sensu* Wolfe & Upchurch 1987), which reflects restructuring of ecosystems following extinctions caused by catastrophic environmental perturbations. Scale of environmental disturbance is one of the most important but also one of the most difficult palaeoecological criteria to assess.

Scale is similarly important in vertical sampling. For example, the closely-spaced sampling of spore assemblages through coals by Bartram (1987a, b) revealed small-scale changes in diversity that would not have been detected by coarser sampling regimes. More importantly, the work also demonstrated that vegetational changes do not correspond well with the lithological boundaries that are generally used to delimit individual samples. Certainly, sampling that is too coarse can deleteriously mix communities just as effectively as taphonomic factors. Detailed sampling also allows comparison of within-horizon and between-horizon dominance-diversity (e.g. Fig. 2.13). The relative abundance of a species within a horizon is treated as a direct measure of its dominance, but it is its frequency *among* horizons that more accurately reflects its ecological role. In all but the most persistently disturbed environments, edaphic climax vegetation is likely to characterize more plant-bearing horizons than are earlier successional seres. For these reasons, and to assist whole-plant reconstruction, I believe that sampling of horizons a few centimetres thick was essential to the success of many of the above case studies and should be encouraged.

In addition to scale of sampling, the relationship between scale of communities and scale of habitat should also be considered. To a Silurian cooksonioid, a few centimetres tall and anchored only by rhizoids, a molehill would be the equivalent of a mountain to a 40-m-tall arboreous lycopsid from the Carboniferous, both in terms of topographical scale and difficulty of colonization. Thus, a Devonian swamp microhabitat (such as those apparent in the Rhynie Chert) may in many ways be the most valid unit of comparison with a Carboniferous macrohabitat (such as those observed in the Herrin Coal). Also, as the scale of the average taphonomic catchment is unlikely to have increased during the Palaeozoic (certainly not to the same degree as the plants), the risk that allochthonous assemblages will contain plant debris from more than one community will on average be proportionately greater in earlier assemblages.

2.10.11 Integration and collaboration

More by historical accident than design, two distinct approaches to the study of

Palaeozoic palaeoecology have evolved. The first is the characterization of specific plant-bearing localities; the tremendous ranges of materials and techniques available at this level are well reflected in the excellent case studies outlined above. Although comparison among them reveals weaknesses, a project combining the greatest strengths of each case study would be (and hopefully will be) superlative. The second approach is 'palaeofloristic analysis.' Traditionally, this consisted primarily of treatise compilation; the amalgamation of taxonomic lists from locality studies to determine the spatial and temporal distributions of taxa. Increasingly, compilation of such lists (and exploitation of pre-existing lists) has been driven by more biological questions requiring assessment of biodiversity and/or community composition (a good example is the study by Knoll (1986) reviewed in section 2.10.7). These synthetic studies require exceptional experience and expertise to yield optimal results. This dichotomy of approaches makes interpretational sense; indeed, I would argue that it should be enhanced. Generalized studies encompassing broad geographical and/or temporal spans and a wide range of quantity and quality of palaeobotanical data are the best means of identifying where and when specific evolutionary events (notably the origin of a major clade and its subsequent radiation) took place. They provide a broad framework for palaeoecological studies but they cannot address the key questions of how and why species from different clades became integrated into recognizable communities occupying recognizable habitats. Detailed understanding of Palaeozoic community composition, structure and dynamics requires detailed, fully integrated palaeontological and sedimentological studies of individual localities. Much time and effort would be consumed during a hypothetical 'perfect' palaeoecological investigation, particularly if it incorporated the highly desirable complementary procedures of whole-plant reconstruction and quantitative floristic analysis. Consequently, study localities should be chosen with great care. The ideal locality is well exposed, spatially and temporally extensive, and encompasses large numbers of assemblages that contain an authochthonous and parachthonous mix of well-preserved, well-articulated mega- and microfossils; such assemblages minimize the need for taphonomic inference. The collection of fresh primary data at appropriately fine scales from the best localities would enhance synthetic studies, especially if conducted in collaboration by specialists familiar with specific localities and taxa.

ACKNOWLEDGEMENTS

I have benefited greatly from access to the publications (past and future) and verbal opinions of members of the Smithsonian Institution's 'Evolution of Terrestrial Ecosystems Consortium,' notably W. A. DiMichele and S. L. Wing. I am grateful to them, together with O. S. Farrington, for their comments on the manuscript, while absolving them from my failure to accept all of the advice offered. I also thank C. J. Cleal for his encouragement and, above all, patience. M. Parrish kindly prepared final drafts of some of the figures, and T. L. Phillips and A. C. Scott generously consented to the reproduction of figures originally published by them. I acknowledge the fiscal support of a Smithsonian Institution Post-Doctoral Research Fellowship.

REFERENCES

Andrews, H. N., Gensel, P. G. & Forbes, W. H. (1974) An apparently heterosporous plant from the Middle Devonian of New Brunswick. *Palaeontology* **17** 387–408.

Andrews, H. N., Kaspar, A., Forbes, W. H., Gensel, P. G. & Chaloner, W. G. (1977) Early Devonian flora of the Trout Valley Formation of northern Maine. *Rev. Palaeobot. Palynol.* **23** 255–285.

Arthur, W. (1984) *Mechanisms of morphological evolution.* Wiley, New York.

Banks, H. P. (1968) The early hisory of land plants. In: Drake, E. T. (ed.), *Evolution and environment.* Yale University Press, New Haven, pp. 73–107.

Banks, H. P. (1975) Reclassification of the Psilophyta. *Taxon* **24** 401–413.

Banks, H. P. (1980) Floral assemblages in the Siluro-Devonian. In: Dilcher, D. L. & Taylor, T. N. (eds.), *Biostratigraphy of fossil plants.* Dowden, Hutchinson & Ross, Stroudsburg, PA, pp. 1–24.

Bartram, K. M. (1987a) *The palynology and petrology of the Barnsley Seam, Westphalian B, Yorkshire, England.* PhD thesis, London University.

Bartram, K. M. (1987b) Lycopod succession in coals: an example from the Low Barnsley Seam (Westphalian B), Yorkshire, England. In: Scott, A. C. (ed.), *Coal and coal-bearing strata: Recent advances. Geol. Soc. Lond. Spec. Pub.* **32**, 187–199.

Bateman, R. M. (1988) *Palaeobotany and palaeoenvironments of Lower Carboniferous floras from two volcanigenic terrains in the Scottish Midland Valley.* PhD thesis, London University.

Bateman, R. M. (1989) Interpretation of heavy mineral assemblage: Outmoded art or undervalued science? *28th Int. Geol. Congr. Abstr.* **1** 97–98.

Bateman, R. M. (1990) The relationship between formal and informal nomenclature and phylogeny in higher taxa: a pedant's perspective on the lycopsids. *Taxon* **39** 624–629.

Bateman, R. M. (in press) Palaeobiological and phylogenetic implications of anatomically-preserved archaeocalamites from the Dinantian of Oxroad Bay and Loch Humphrey Burn, Scotland. *Palaeontographica B* **214**.

Bateman, R. M. & Denholm, I. (1989) The complementary roles of organisms, populations and species in 'demographic' phytosystematics. *Amer. J. Bot.* **76** 226.

Bateman, R. M. & Rothwell, G. W. (1990) A reappraisal of the Dinantian floras at Oxroad Bay, East Lothian, Scotland. 1. Floristics and the development of whole-plant concepts. *Trans. R. Soc. Edinburgh B* **81** 127–159.

Bateman, R. M. & Scott, A. C. (1990) A reappraisal of the Dinantian floras at Oxroad Bay, East Lothian, Scotland. 2. Volcanicity, palaeoecology and palaeoenvironments. *Trans. R. Soc. Edinburgh B* **81** 161–194.

Bateman, R. M., DiMichele, W. A. & Willard, D. A. (in press) Experimental cladistic analyses of anatomically-preserved arborescent lycopsids from the Carboniferous of Euramerica: an essay in paleobotanical phylogenetics. *Ann. Mo. Bot. Gard.*

Batenburg, L. H. (1982) 'Compression species' and 'petrifaction species' of *Sphenophyllum* compared. *Rev. Palaeobot. Palynol.* **36** 335–359.

Beck, C. B. (1964) Predominance of *Archaeopteris* in Upper Devonian floras of western Catskills and adjacent Pennsylvania. *Bot. Gaz.* **125** 126–128.

Beck, C. B., Coy, K. & Schmid, R. (1982) Observations on the fine structure of *Callixylon* wood. *Amer. J. Bot.* **69** 54–76.

Beckett, H. (1845) On a fossil forest in the Parkfield Colliery, near Wolverhampton. *Quart. J. Geol. Soc.* **1** 41–43.

Beerbower, J. R. (1985) Early development of continental ecosystems. In: Tiffney, B. H. (ed.), *Geological factors and the evolution of plants*. Yale University Press, New Haven, pp. 47–91.

Behrensmeyer, A. K. & Kidwell, S. M. (1985) Taphonomy's contributions to paleobiology. *Paleobiology* **11** 105–119.

Behrensmeyer, A. K., Hook, R. W., Boy, J. A., Dodson, P., Gastaldo, R. A., Spicer, R. A., Wilson, M. V. H., Graham, R. W., Badgley, C. E. & Olsen, P. E. (in press) Paleoenvironmental contexts and taphonomic modes in the terrestrial fossil record. In ETE Consortium (eds.), *Evolutionary paleoecology of plants and animals*. Chicago University Press, Chicago.

Bott, M. H. P. (1987) Subsidence mechanisms of Carboniferous basins in northern England. In: Miller, J., Adams, A. E. & Wright, V. P. (eds.), *European Dinantian environments*. Wiley, New York, pp. 21–32.

Briggs, D. J. (1977) *Sources and methods in geography: sediments*. Butterworth, London.

Brooks, D. R. & McClennan, D. A. (1991) *Phylogeny, ecology and behavior*. Chicago University Press, Chicago.

Burnham, R. J. (1989) Relationship between standing vegetation and leaf litter in a paratropical forest: implications for paleobotany. *Rev. Palaeobot. Palynol.* **58** 5–32.

Burnham, R. J., Parker, G. G. & Wing, S. L. (in press) Forest litter accurately reflects stand composition. *Paleobiology* **17**.

Chaloner W. G. (1986) Reassembling the whole plant, and naming it. In: Spicer, R. A. & Thomas, B. A. (eds.), *Systematic and taxonomic approaches in palaeobotany. Syst. Ass. Spec. Vol. 31*. University Press, Oxford, pp. 67–78.

Chaloner, W. G. (1988) Early land plants: the saga of a great conquest. In Greuter, W. & Zimmer, B. (eds.), *Proc. 14th. Int. Bot. Congr.* pp. 301–316. Koeltz, Königstein.

Chaloner, W. G. & Gay, M. M. (1973) Scanning electron microscopy of latex casts of fossil plant impressions. *Palaeontology* **16** 645–649.

Chaloner, W. G. & Lawson, J. D. (eds.) (1985) Evolution and environment in the late Silurian and early Devonian. *Phil. Trans. R. Soc. Lond. B* **309** 1–342.

Chaloner, W. G. & MacDonald, P. (1980) *Plants invade the land*. H.M.S.O./Royal Scottish Museum, Edinburgh.

Chaloner, W. G. & Sheerin, A. (1979) Devonian macrofloras. In: House, M. R. *et al*. (eds.), *The Devonian System. Spec. Pap. Palaeont.* **23** 145–161.

Chapman, D. J. (1985) Geological factors and biochemical aspects of the origin of land-plants. In: Tiffney, B. H. (ed.), *Geological factors and the evolution of plants.* Yale University Press, New Haven, pp. 23–45.

Clayton, G., Coquel, R. Doubinger, J., Gueinn, K. J., Loboziak, S., Owens, B. & Streel, M. (1977) Carboniferous miospores of western Europe: illustration and zonation. *Meded. Rijks. Geol. Dienst.* **29** 1–71.

Cleal, C. J. (1987) This is the forest primaeval. *Nature* **326** 828.

Cleal, C. J. & Zodrow, E. L. (1989) Epidermal structure of some medullosan *Neuropteris* foliage from the Middle and Upper Carboniferous of Canada and Germany. *Palaeontology* **32** 837–882.

Collinson, M. E. & Scott, A. C. (1987a) Implications of vegetational change through the geological record on models of coal-forming environments. In: Scott, A. C. (ed.), *Coal and coal-bearing strata: recent advances. Geol. Soc. Lond. Spec. Pub. 32.* Blackwell, London, pp. 67–85.

Collinson, M. E. & Scott, A. C. (1987b) Factors controlling the organisation and evolution of ancient plant communities. In: Gee, G. H. R. & Giller, P. S. (eds.), *Organisation of communities past and present.* Blackwell, Oxford, pp. 399–420.

Cope, M. J. & Chaloner, W. G. (1985) Wildfire: An interaction of biological and physical processes. In: Tiffney, B. H. (ed.), *Geological factors and the evolution of plants.* Yale University Press, New Haven, pp. 257–277.

Costanza, S. (1983) *Morphology and systematics of cordaites of Pennsylvanian coal swamps in Euramerica.* PhD thesis, University of Illinois.

Courvoisier, J. M. & Phillips, T. L. (1975) Correlation of spores from Pennsylvanian coal-ball fructifications with dispersed spores. *Micropaleont.* **21** 45–59.

Crane, P. R. (1989) Patterns of evolution and extinction in vascular plants. In: Allen, K. C. & Briggs, D. E. G. (eds.), *Evolution and the fossil record.* Belhaven, Chichester, pp. 153–187.

Creber, G. T. & Chaloner, W. G. (1984) Influence of environmental factors on the wood structure of living and fossil trees. *Bot. Rev.* **50** 357–448.

Crowson, R. A. (1970) *Classification and biology*. Heinemann, London.

Cutler, D. F., Alvin, K. L. & Price, C. E. (eds.) (1982) *The plant cuticle.* Linn. Soc. Symp. Ser. 10. London. 461 pp.

Dahlquist, R. L. & Knoll, J. W. (1978) Inductively-coupled plasma-atomic emission spectrometry analysis of biological materials and soils for major, trace and ultra-trace elements. *Appl. Spectrosc.* **32** 1–30.

Darrah, W. C. (1969) *A critical review of the Upper Pennsylvanian floras of Eastern United States with notes on the Mazon Creek flora of Illinois.* W. C. Darrah, Gettysburg, Pennsylvania.

Davies, D. (1929) Correlation and palaeontology of the Coal Measures in East Glamorganshire. *Phil. Trans. R. Soc. Lond. B* **217** 91–154.

Delevoryas, T. (1955) The Medullosae: structure and relationships. *Palaeontographica B* **97** 114–167.

DeMaris, P. J., Bauer, R. A., Cahill, R. A. & Damberger, H. H. (1983) Geologic investigation of roof and floor strata: longwall demonstration, Old Ben Mine No. 24; prediction of coal balls in the Herrin Coal. *Illinois State Geol. Surv. Rep. 1983/2.* 69 pp.

Digby, P. G. N. & Kempton, R. A. (1987) *Multivariate analysis of ecological communities*. Chapman & Hall, London.

DiMichele, W. A. (1985) *Diaphorodendron*, gen. nov., a segregate from *Lepidodendron (Pennsylvanian-age)*. *Syst. Bot.* **10** 453–458.

DiMichele, W. A. & Aronson, R. B. (in press) The Pennsylvanian–Permian vegetation transition: a terrestrial analogue to the onshore–offshore hypothesis. *Evolution* **45**.

DiMichele, W. A. & Bateman, R. M. (in press) Diaphorodendraceae, fam. nov. (Carboniferous): implications of *Diaphorodendron* and *Synchysidendron*, gen. nov., for lycopsid evolution and classification. *Amer. J. Bot.* **78**.

DiMichele, W. A. & DeMaris, P. J. (1987) Structure and dynamics of a Pennsylvanian-age *Lepidodendron* forest: colonizers of a disturbed swamp habitat in the Herrin (No. 6) Coal of Illinois. *Palaios* **2** 146–157.

DiMichele, W. A. & Nelson, W. J. (1989) Small-scale spatial heterogeneity in Pennsylvanian-age vegetation from the roof shale of the Springfield Coal (Illinois Basin). *Palaios* **4** 276–280.

DiMichele, W. A. & Phillips, T. L. (1985) Arborescent lycopod reproduction and paleoecology in a coal-swamp environment of late Middle Pennsylvanian age (Herrin Coal, Illinois, U.S.A.). *Rev. Palaeobot. Palynol.* **44** 1–26.

DiMichele, W. A. & Phillips, T. L. (1988) Paleoecology of the Middle Pennsylvanian-age Herrin Coal swamp (Illinois) near a contemporaneous river system, the Walshville paleochannel. *Rev. Palaeobot. Palynol.* **56** 151–176.

DiMichele, W. A., Phillips, T. L. & Peppers, R. A. (1985) The influence of climate and depositional environment on the distribution and evolution of Pennsylvanian coal-swamp plants. In: Tiffney, B. H. (ed.), *Geological factors and the evolution of plants*. Yale University Press, New Haven, pp. 223–256.

DiMichele, W. A., Phillips, T. L. & Willard, D. A. (1986) Morphology and palaeoecology of Pennsylvanian-age coal-swamp plants. In: Gastaldo, R. A. & Broadhead, T. W. (eds.), *Land plants: notes for a short course*. Univ. Tennessee Dept. Geol. Sci. Stud. Geol. 15, pp. 97–114.

DiMichele, W. A., Phillips, T. L. & Olmstead, R. G. (1987) Opportunistic evolution: abiotic environmental stress and the fossil record of plants. *Rev. Palaeobot. Palynol.* **50** 151–178.

DiMichele, W. A., Davis, J. I. & Olmstead, R. G. (1989) Origins of heterospory and the seed habit. *Taxon* **38** 1–11.

DiMichele, W. A., Hook, R. W., Beerbower, J. R., Boy, J., Gastaldo, R. A., Hotton, N. III, Phillips, T. L., Scheckler, S. E., Shear, W. A. & Sues, H.-D. (in press) Palaeozoic terrestrial ecosystems: review and overview. In ETE Consortium (eds.), *Evolutionary paleoecology of plants and animals* Chicago University Press, Chicago.

Donoghue, M. J. (1989) Phylogenies and the analysis of evolutionary sequences, with examples from seed plants. *Evolution* **43** 1137–1156.

Drägert, K. (1964) Pflanzensoziologische untersuchungen in den Mittleren Essener Schichten des nördlichen Ruhrgebietes. *Forsch. Ber. Landes N. Rhein-Westf.* **1363** 1–295.

Edwards, D. (1980) Early land floras. In: Panchen, A. R. (ed.), *The terrestrial environment and the origin of land vertebrates*. Systematics Association Spec. Vol. 15. Academic Press, London, pp. 55–85.

Edwards, D. (1982) Fragmentary non-vascular plant microfossils from the late Silurian of Wales. *Bot. J. Linn. Soc.* **84** 223–256.

Edwards, D. & Edwards, D. S. (1986) A reconsideration of the Rhyniophytina Banks. In: Spicer, R. A. & Thomas, B. A. (eds.), *Systematic and taxonomic approaches in palaeobotany. Systematics Association Spec. Vol. 31.* University Press, Oxford, pp. 199-220.

Edwards, D. & Fanning, U. (1985) Evolution and environment in the late Silurian–early Devonian: the rise of the pteridophytes. *Phil. Trans. R. Soc. Lond. B* **309** 147–165.

Edwards, D., Feehan, J. & Smith, D. G. (1983) A Late Wenlock flora from Co. Tipperary, Ireland. *Bot. J. Linn. Soc.* **86** 19–36.

Edwards, D. S. (1980) Evidence for the sporophyte status of the Lower Devonian plant *Rhynia gwynne-vaughanii* Kidston & Lang. *Rev. Palaeobot. Palynol.* **29** 177–188.

Edwards, D. S. (1986) *Aglaophyton major*, a non-vascular plant from the Devonian Rhynie Chert. *Bot. J. Linn. Soc.* **93** 173–204.

Edwards, D. S. & Lyon, A. G. (1983) Algae from the Rhynie Chert. *Bot. J. Linn. Soc.* **86** 37–55.

Efremov, I. A. (1940) Taphonomy: a new branch of paleontology. *Pan-Am. Geol.* **74** 81–93.

Eggert, D. A. (1974) The sporangium of *Horneophyton lignieri* (Rhyniophytina). *Amer. J. Bot.* **61** 405–413.

Eldredge, N. (1989) *Macroevolutionary dynamics: species, niches, and adaptive peaks*. McGraw-Hill, New York.

El-Saadaway, W. E. & Lacey, W. S. (1979a) Observations on *Nothia aphylla* Lyon ex Hoeg. *Rev. Palaeobot. Palynol.* **27** 119–147.

El-Saadaway, W. E. & Lacey, W. S. (1979b) The sporangia of *Horneophyton lignieri* (Kidston & Lang) Barghoorn & Darrah. *Rev. Palaeobot. Palynol.* **28** 137–144.

Evolution of Terrestrial Ecosystems Consortium (eds.). (in press) *Evolutionary paleoecology of plants and animals*. Chicago University Press, Chicago.

Fairon-Demaret, M. (1986) Some uppermost Devonian megafloras: a stratigraphical review. *Ann. Soc. géol. Belg.* **109** 43–48.

Fanning, U., Richardson, J. B. & Edwards, D. (1988) Cryptic evolution in an early land plant. *Evol. Tr. Pl.* **2** 13–24.

Farley, M. B. (1988) Environmental variation, palynofloras, and paleoecological interpretation. In: DiMichele, W. A. & Wing, S. L. (eds.), *Methods and applications of plant paleoecology*. Paleont. Soc. Spec. Pub. 3, pp. 126–146.

Feakes, C. R. & Retallack, G. J. (1988) Recognition and chemical characterization of fossil soils developed on alluvium: a Late Ordovician example. *Geol. Soc. Amer. Spec. Pap.* **216** 35–48.

Ferguson, D. K. (1985) The origin of leaf assemblages: new light on an old problem. *Rev. Palaeobot. Palynol.* **46** 117–188.

Fritz, W. J. (1980) Stumps transported and deposited upright by Mt. St. Helens mudflows. *Geology* **8** 568–588.

Fritz, W. J. (1986) Plant taphonomy in areas of explosive volcanism. In: Gastaldo, R. A. & Broadhead, T. W. (ed.), *Land plants: notes for a short course*. Univ. Tennessee Dept. Geol. Sci. Stud. Geol. 15, pp. 1–9.

Fritz, W. J. & Harrison, S. (1985) Transported trees from the 1982 Mount St. Helens sediment flows: their use as paleocurrent indicators. *Sedim. Geol.* **42** 49–64.

Galtier, J. (1986) Taxonomic problems due to preservation: comparing compression and permineralised taxa. In: Spicer, R. A. & Thomas, B. A. (eds.), *Systematic and taxonomic approaches in palaeobotany. Syst. Ass. Spec. Vol. 31.* Oxford University Press, Oxford, pp. 1–16.

Galtier, J. & Phillips, T. L. (1985) Swamp vegetation from Grand Croix (Stephanian) and Autun (Autunian), France, and comparison with coal-ball peats of the Illinois Basin. *C. R. 9e Congr. Inst. Strat. Géol. Carbon.* (Washington & Urbana, 1979) **4** 13–24.

Galtier, J. & Scott, A. C. (1985) The diversification of early ferns. *Proc. R. Soc. Edinburgh B* **86** 289–301.

Galtier, J. & Scott, A. C. (1990) On *Eristophyton* and other gymnosperms from the Lower Carboniferous of Castleton Bay, East Lothian, Scotland. *Geobios* **23** 5–19.

Garratt, M. J. & Rickards, R. B. (1984) Graptolite biostratigraphy of early land plants from Victoria, Australia. *Proc. Yorks. Geol. Soc.* **44** 377–384.

Gastaldo, R. A. (1984) A case against pelagochthony: the untenability of Carboniferous arborescent lycopod-dominated floating peat mats. In: K. R. Walker (ed.), *The evolution/creation controversy*. Paleont. Soc. Spec. Pap. 1, pp. 97–116.

Gastaldo, R. A. (1985a) Plant accumulating deltaic depositional environments: Mobile Delta, Alabama. *Alabama Geol. Surv. Repr. Ser.* **66** 1–35.

Gastaldo, R.A. (1985b) Upper Carboniferous paleoecological reconstructions: observations and reconsiderations. *C. R. 10e Congr. Int. Strat. Gol. Carbon.* (Madrid, 1983) **2** 281–296.

Gastaldo, R. A. (1986a) Implications on the paleoecology of autochthonous lycopods in clastic sedimentary environments of the Early Pennsylvanian of Alabama. *Palaeogeog. Palaeoclim. Palaeoecol.* **53** 191–212.

Gastaldo, R. A. (1986b) An explanation for lycopod configuration, 'Fossil Grove', Victoria Park, Glasgow. *Scott. J. Geol.* **22** 77–83.

Gastaldo, R. A. (1987) Confirmation of Carboniferous clastic swamp communities. *Nature* **326** 869–871.

Gastaldo, R. A. (1988) Conspectus of phytotaphonomy. In: DiMichele, W. A. & Wing, S. L. (eds.), *Methods and applications of plant paleoecology*. Paleont. Soc. Spec. Pub. 3, pp. 14–28.

Gastaldo, R. A. (1990) The paleobotanical character of log assemblages necessary to differentiate blow-downs resulting from cyclonic winds. *Palaios* **5** 472–478.

Gastaldo, R. A. & Broadhead, T. W. (eds.) (1986) *Land plants: notes for a short course*. Univ. Tennessee Dept. Geol. Sci. Stud. Geol. 15.

Gensel, P. G. (1986) Diversification of land plants in the Early and Middle

Devonian. In: Gastaldo, R. A. & Broadhead, T. W. (eds.), *Land plants: Notes for a short course*. Univ. Tennessee Dept. Geol. Sci. Stud. Geol. 15, pp. 64–80.

Gensel, P. G. (in press) The relationships of the zosterophylls and lycopods: Evidence from morphology, paleoecology and cladistic methods of inference. *Ann. Mo. Bot. Gard.*

Gensel, P. G. & Andrews, H. N. (1984) *Plant life in the Devonian*. Praeger, New York.

Gingerich, P. D. (1979) Stratophenetic approach to phylogeny reconstruction in vertebrate paleontology. In: Cracraft, J. & Eldredge, N. (eds.), *Phylogenetic analysis and paleontology*. Columbia University Press, New York. pp. 41–79.

Good, C. W. (1975) Pennsylvanian-age calamitean cones, elater-bearing spores, and associated vegetative organs. *Palaeontographica B* **153** 28–99.

Gordon, W. T. (1909) On the nature and occurrence of the plant-bearing rocks at Pettycur, Fife. *Trans. Edinburgh Geol. Soc.* **9** 355–360.

Gordon, W. T. (1935) The genus *Pitys* Witham emend. *Trans. R. Soc. Edinburgh* **58** 279–311.

Gordon, W. T. (1938) On *Tetrastichia bupatides*: a Carboniferous pteridosperm from East Lothian. *Trans. R. Soc. Edinburgh* **59** 351–370.

Grant, P. R. (1986) Time and evolution. In: Spicer, R. A. & Thomas, B. A (eds.), *Systematic and taxonomic approaches in palaeobotany*. *Syst. Ass. Spec. Vol. 31*. University Press, Oxford, pp. 263–274.

Gray, J. (1984) Ordovician–Silurian land plants: The interdependence of ecology and evolution. *Spec. Pap. Palaeont.* **32** 281–295.

Gray, J. (1985) The microfossil record of early land plants: advances in understanding of early terrestrialization, 1970–1984. *Phil. Trans. R. Soc. London B* **309** 167–195.

Grayson, R. F. & Oldham, L. (1987) A new structural framework for the northern British Dinantian as a basis for oil, gas and mineral exploration. In: Miller, J., Adams, A. E. & Wright, V. P. (eds.), *European Dinantian Environments*. Wiley, New York, pp. 33–59.

Grierson, J. D. & Banks, H. P. (1963) Lycopods of the Devonian of New York State. *Palaeontographica Amer.* **4** 220–279.

Grime, J. P. (1979) *Plant strategies and vegetation processes*. Wiley, New York.

Harland, W. B., Armstrong, R. V., Cox, A. V., Craig, L. E., Smith, A. G. & Smith, D. G. (1989) *A geologic time scale 1989.* Cambridge University Press, Cambridge.

Harper, J. L. (1977) *The population biology of plants*. Academic Press, London.

Harris, T. M. (1981) Burnt ferns in the English Wealden. *Proc. Geol. Ass.* **92** 47–58.

Havlena, V. (1970) Einige Bermukungen zur Phytogeographie und Geobotanik des Karbons und Perms. *C. R. 6e Congr. Inst. Strat. Gol. Carbon*. (Sheffield, 1967) **3** 901–912.

Hill, C. R. (1986) The epidermis/cuticle and *in situ* spores and pollen in fossil plant taxonomy. In: Spicer, R. A. & Thomas, B. A. (eds.), *Systematic and taxonomic approaches in palaeobotany*. *Syst. Ass. Spec. Vol. 31*. University Press, Oxford, pp. 123–136.

Hirmer, M. (1927) *Handbuch der Paläobotanik*. Oldenbourg, Munich.

Holmes, P. L. (in press) Differential transport of spores and pollen: a laboratory study. *Rev. Palaeobot. Palynol.*

Horne, J. *et al.* (1916) The plant-bearing cherts of Rhynie, Aberdeenshire. *Br. Ass. Adv. Sci. Rep. 1916* 206–216.

Hueber, F. M. (in press) Speculations on the early lycopsids and zosterophylls. *Ann. Mo. Bot. Gard.*

Hutchinson, G. E. (1965) The niche: an abstractly inhabited hypervolume. In: Hutchinson, G. E. (ed.), *The ecological theater and the evolutionary play*. Yale University Press, New Haven, pp. 26–78.

Jennings, J. R. (1976) The morphology and relationships of *Rhodea, Telangium, Telangiopsis* and *Heterangium. Amer. J. Bot.* **63** 1119–1133.

Johnson, D. O. (1972) *Stratigraphic analysis of the interval between the Herrin (No. 6) Coal and the Piasa Limestone in southwestern Illinois*. PhD thesis, University of Illinois.

Johnson, G. A. L. & Tarling, D. H. (1985) Continental convergence and closing seas during the Carboniferous. *C. R. 10e Congr. Inst. Strat. Gol. Carbon*. (Madrid, 1983) **4** 163–168.

Joy, K. W., Willis, A. J. & Lacey, W. S. (1956) A rapid cellulose peel technique in palaeobotany. *Ann. Bot. (NS)* **20** 635–637.

Kenrick, P. & Crane, P. R. (1990) Cell wall ultrastructure in early land plants: phylogenetic implications (abstr.). *Amer. J. Bot.* **77** 140.

Kenrick, P. & Crane, P. R. (in press) Water-conducting cells of *Rhynia gwynne-vaughanii* and *Asteroxylon mackiei* from the Rhynie Chert (Lower Devonian): implications for the early evolution of land plants. *Bot. Gaz.*

Kenrick, P. & Edwards, D. (1988) The anatomy of Lower Devonian *Gosslingia breconensis* Heard based on pyritized axes, with some comments in the permineralization process. *Bot. J. Linn. Soc.* **97** 95–123.

Kenrick, P, Remy, W. & Crane, P. R. (1991) The structure of water conducting cells in the enigmatic early land plants *Stockmansella langii* Fairon-Demaret, *Huvenia kleui* Hass & Remy, and *Sciadophyton* sp. Remy *et al.* 1980. *Argum. Palaeobot.* **8** 179–191.

Kerp, J. H. F. & Fichter, J. (1985) Die Makrofloren des Saarpfalzischen Rotliegenden (Ober Karbon-Unter Perm; SW Deutschland). *Mainz. geowiss. Mitt.* **14** 159–286.

Kerp, J. H. F., Poort, R., Swinkels, H. & Verwer, R. (1989) A conifer-dominated flora from the Rotliegend of Oberhausen (Saar-Nahe area). *Cour. Forsch.-Inst. Senckenberg* **109** 137–151.

Kevan, P. G., Chaloner, W. G. & Saville, D. B. O. (1975) Interrelationships of early terrestrial arthropods and plants. *Palaeontology* **19** 391–417.

Kidston, R. & Lang, W. H. (1917) On Old Red Sandstone plants showing structure from the Rhynie Chert Bed, Aberdeenshire. I. *Rhynia gwynne-vaughanii* Kidston & Lang. *Trans. R. Soc. Edinburgh B* **51** 761–784.

Kidston, R. & Lang, W. H. (1920a) On Old Red Sandstone plants showing structure from the Rhynie Chert Bed, Aberdeenshire. II. Additional notes on *Rhynia major* n. sp. and *Hornea lignieri* n. g., n. sp. *Trans. R. Soc. Edinburgh B* **52** 603–627.

Kidston, R. & Lang, W. H. (1920b) On Old Red Sandstone plants showing structure from the Rhynie Chert Bed, Aberdeenshire. III. *Asteroxylon mackiei* Kidston & Lang. *Trans. R. Soc. Edinburgh B* **52** 643–680.

Kidston, R. & Lang, W. H. (1921a) On Old Red Sandstone plants showing structure from the Rhynie Chert Bed, Aberdeenshire. IV. Restorations of the vascular cryptogams, and discussion of their bearing on the general morphology of Pteridophyta and the origin of the organisation of land plants. *Trans. R. Soc. Edinburgh B* **52** 831–854.

Kidston, R. & Lang, W. H. (1921b) On Old Red Sandstone plants showing structure from the Rhynie Chert Bed, Aberdeenshire. V. The Thallophyta occurring in the peat bed, the succession of plants occurring throughout a vertical section of the bed, and the conditions of accumulation and preservation of the bed. *Trans. R. Soc. Edinburgh B* **52** 855–902.

Knoll, A. H. (1984) Patterns of extinction in the fossil record of vascular plants. In Nitecki, M. (ed.), *Extinctions* 21–68. Chicago University Press, Chicago.

Knoll, A. H. (1985a) Exceptional preservation of photosynthetic organisms in silicified carbonates and silicified peats. *Phil. Trans. R. Soc. Lond. B* **311** 111–122.

Knoll, A. H. (1985b) The distribution and evolution of microbial life in the Late Proterozoic Era. *Ann. Rev. Microbiol.* **39** 391–417.

Knoll, A. H. (1986) Patterns of change in plant communities through geological time. In Diamond, J. & Case, T. J. (eds.), *Community ecology*. Harper & Row, New York, pp. 126–141.

Knoll, A. H. (1987) Protists and Phanerozoic evolution in the oceans. In: Broadhead, T. W. & Lipps, J. (eds.), *Fossil prokaryotes and protists: notes for a short course 2*. Univ. Tennessee Dept. Geol. Sci. Stud. Geol. 18, pp. 248–264.

Knoll, A. H. & Niklas, K. J. (1987) Adaptation, plant evolution, and the fossil record. *Rev. Palaeobot. Palynol.* **50** 127–149.

Knoll, A. H., Niklas, K. J. & Tiffney, B. H. (1979) Phanerozoic land plant diversity in North America. *Science* **206** 1400–1402.

Knoll, A. H., Niklas, K. J., Gensel, P. G. & Tiffney, B. H. (1984) Character diversification and patterns of evolution in early vascular plants. *Paleobiology* **10** 34–47.

Knoll, A. H., Grant, S. W. F. & Tsao, J. W. (1986) The early evolution of land plants. In: Gastaldo, R. A. & Broadhead, T. W. (ed.), *Land plants: notes for a short course*. Univ. Tennessee Dept. Geol. Sci. Stud. Geol. 15. pp. 45–63.

Kovach, W. L. (1988) Multivariate methods of analyzing paleoecological data. In: DiMichele, W. A. & Wing, S. L. (eds.), *Methods and applications of plant paleoecology*. Paleont. Soc. Spec. Pub. 3, pp. 72–104.

Krassilov, V. A. (1975) *Paleoecology of terrestrial plants: basic principles and techniques*. Wiley, New York.

Kryshtofovich, A. N. (1944) The mode of preservation of plant fossils and its bearing on the problem of coal. *Amer. J. Sci.* **242** 57–73.

Kryshtofovich, A. N. (1957) *Palaeobotany*. Gostoptekhizdat, Leningrad [In Russian).

Kummel, B. & Raup, D. (1965) *Handbook of paleontological techniques*. Freeman, San Francisco.

Labandeira, C. C. & Beall, B. S. (1990) Arthropod terrestriality. In: Mikulic, S. J. (ed.), *Arthropod paleobiology*. Paleont. Soc. Short Cour. Paleont. 3, pp. 214–256.

Lang, W. H. (1937) On the plant remains from the Downtonian of England and Wales. *Phil. Trans. R. Soc. Lond. B* **227** 245–291.

Leary, R. L. (1981) Early Pennsylvanian geology and palaeobotany of the Rock Island County, Illinois area, 1. Geology. *Illinois State Mus. Rep. Inv.* **37** 88 pp.

Leary, R. L. & Pfefferkorn, H. W. (1977) An early Pennsylvanian flora with *Megalopteris* and Noeggerathiales from west-central Illinois. *Illinois State Geol. Surv. Circ. 500*. 77 pp.

Leeder, M. R. (1987) Plate tectonics, palaeogeography and sedimentation in Lower Carboniferous Europe. In: Miller, J., Adams, A. E. & Wright, V. P. (eds.), *European Dinantian environments* Wiley, New York, pp. 1–20.

Leeder, M. R. (1988) Recent developments in Carboniferous geology: a critical review with implications for the British Isles and NW Europe. *Proc. Geol. Ass.* **99** 73–100.

Lemoigne, Y. (1968) Observation d'archégones ports par des axes du type *Rhynia gwynne-vaughanii* Kidston et Lang. Existence de gametophytes vascularisés au Devonien. *C. R. Acad. Sci. Paris* **266** 1655–1657.

Leo, R. F. & Barghoorn, E. S. (1976) Silicification of wood. *Bot. Mus. Leaflts Harvard Univ.* **25** 1–46.

Lesnikowska, A. D. (1989) *Anatomically-preserved Marattiales from coal swamps of the Desmoinsean and Missourian of the midcontinent United States: systematics, ecology, and evolution*. PhD thesis, University of Illinois.

Leys, C. A. (1982) *Volcanic and sedimentary processes in phreatomagmatic volcanoes*. PhD thesis, Leeds University.

Loftus, G. W. F. & Greensmith, J. T. (1988) The lacustrine Burdiehouse Limestone Formation: a key to the deposition of the Dinantian oil shales of Scotland. In: Fleet, A. J., Kelts, K. & Talbot, M. R. (eds.), *Lacustrine petroleum source rocks*. Blackwell, Oxford, pp. 219–234.

Long, A. G. (1968) Some specimens of *Mazocarpon, Achlamydocarpon* and *Cystosporites* from the Lower Carboniferous rocks of Berwickshire. *Trans. R. Soc. Edinburgh B* **67** 359–372.

Long, A. G. (1975) Further observations on some Lower Carboniferous seeds and cupules. *Trans. R. Soc. Edinburgh B* **69** 26--293.

Long, A. G. (1979) Observations on the Lower Carboniferous genus *Pitus* Witham. *Trans. R. Soc. Edinburgh* **70** 111–127.

Lyon, A. G. (1964) The probable fertile region of *Asteroxylon mackiei* K. & L. *Nature* **203** 1082–1083.

MacGregor, M. & Walton, J. (1972) *The story of the fossil grove* (3rd edn.). Glasgow District Council, Glasgow.

Mahaffy, J. F. (1985) Profile patterns of peat and coal palynology in the Herrin (No. 6) Coal Member, Corbondale Formation, Middle Pennsylvanian of southern Illinois. *C. R. 9e Congr. Inst. Strat. Géol. Carbon.* (Washington & Urbana, 1979) **5** 25–34.

Mamay, S. H. (1967) Lower Permian plants from the Arroyo Formation in Baylor County, north-central Texas. *U. S. Geol. Surv. Prof. Pap.* **523E.** 15 pp.

Mamay, S. H. (1976) Palaeozoic origin of the cycads. *U. S. Geol. Surv. Prof. Pap.* **934**. 48 pp.

Mamay, S. H. & Bateman, R. M. (1991) *Archaeocalamites lazarii* sp. nov.: the range of the Archaeocalamitaceae extended from the lowermost Pennsylvanian to the mid-Lower Permian. *Amer. J. Bot.* **78** 489–496.

Mapes, G. (1987) Ovule inversion in the earliest conifers. *Amer. J. Bot.* **74** 1205–1210.

Mapes, G. & Gastaldo, R. A. (1986) Late Paleozoic non-peat-accumulating floras. In: Gastaldo, R. A. & Broadhead, T. W. (eds.), *Land plants: notes for a short course*. Univ. Tennessee Dept. Geol. Sci. Stud. Geol. 15, pp. 115–127.

Mapes, G., Rothwell, G. W. & Haworth, M. T. (1989) Evolution of seed dormancy. *Nature* **337** 645–646.

Matten, L. C. (1974) The Givetian flora from Cairo, New York: *Rhacophyton, Triloboxylon* and *Cladoxylon. Bot. J. Linn. Soc.* **68** 303–318.

Matten, L. C., Tanner, W. R. & Lacey, W. S. (1984) Additions to the silicified Upper Devonian/Lower Carboniferous flora from Ballyheigue, Ireland. *Rev. Palaeobot. Palynol.* **43** 303–320.

McComas, M. A. (1988) Upper Pennsylvanian compression floras of the 7–11 Mine, Columbiana County, northeastern Ohio. *Ohio J. Sci.* **88** 48–52.

Meyen, S. V. (1982) The Carboniferous and Permian floras of Angaraland: a synthesis. *Biol..Mem.* **7** 1–110.

Meyen, S. V. (1987) *Fundamentals of palaeobotany*. Chapman & Hall, London.

Millay, M. A. (1979) Studies of Paleozoic marattialeans: a monograph of the American species of *Scolecopteris*. *Palaeontographica B* **169** 1–69.

Mishler, B. D. & Churchill, S. P. (1985) Transition to a land flora: phylogenetic relationships of the green algae and bryophytes. *Cladistics* **1** 305–328.

Moore, P. D. (1987) Ecological and hydrological aspects of peat formation. In: Scott, A. C. (ed.), *Coal and coal-bearing strata: recent advances. Geol. Soc. Spec. Pub.* **32** pp. 7–15.

Murphy, C. P. (1986) *Thin section preparation of soils and sediments*. A B Academic, Berkhamsted.

Neavel, R. C. & Guennel, G. K. (1960) Indiana paper coal: composition and deposition. *J. Sedim. Petrol.* **30** 241–248.

Niklas, K. J. (1976) Morphological and ontogenetic reconstruction of *Parka decipiens* Fleming and *Pachytheca* Hooker from the Lower Old Red Sandstone, Scotland. *Trans. R. Soc. Edinburgh B* **69** 483–499.

Niklas, K. J. (1982) Computer simulations of early land plant branching morphologies: canalization of patterns during evolution? *Paleobiology* **8** 196–210.

Niklas, K. J. (1986) Large-scale changes in animal and plant terrestrial communities. In: Raup, D. M. & Jablonski, D. (eds.), *Patterns and processes in the history of life*. Springer, Berlin, pp. 383–405.

Niklas, K. J. & Banks, H. P. (1990) A reevaluation of the zosterophyllophytina with comments on the origin of lycopods. *Amer. J. Bot.* **77** 274–283.

Niklas, K. J., Tiffney, B. H. & Knoll, A. H. (1980) Apparent changes in the diversity of fossil plants: a preliminary assessment. In: Hecht, M. K., Steere, W. C. & Wallace, B. (eds.), *Evolutionary biology Vol. 12*, Plenum, New York, pp. 1–89.

Niklas, K. J., Tiffney, B. H. & Knoll, A. H. (1983) Patterns in vascular land plant diversification. *Nature* **303** 614–616.

Niklas, K. J., Tiffney, B. H. & Knoll, A. H. (1985) Patterns in vascular land plant diversification: an analysis at the species level. In: Valentine, J. W. (ed.), *Phanerozoic diversity patterns: profiles in macroevolution*, Princeton University Press, Princeton, pp. 97–128.

Odum, E. P. (1971) *Fundamentals of ecology* (3rd edn.). Saunders, Philadelphia.

Oshurkova, M. V. (1974) A facies–paleoecological approach to the study of fossilized plant remains. *Paleont. J.* **3** 363–370.

Pagel, M. D. & Harvey, P. H. (1988) Recent developments in the analysis of comparative data. *Quart. Rev. Biol.* **63** 413–440.

Pant, D. D. (1962) The gametophyte of the Psilophytales. In: Maheshwari, P., Johri, B. M. & Vasil, I. K. (eds.), *Proc. Summer Sch. Bot. (Darjeeling)*, 276–301.

Parrish, J. M., Parrish, J. T. & Ziegler, A. M. (1988) Permian–Triassic paleogeography and paleoclimatology and implications for therapsid distribution. In: Roth, J., Roth, C. & Hotton, N. III (eds.), *The biology and ecology of mammal-like reptiles*. Smithsonian Institution, Washington D.C.

Pfefferkorn, H. W. & Thomson, M. C. (1982) Changes in dominance patterns in Upper Carboniferous plant-fossil assemblages. *Geology* **10** 641–644.

Phillips, T. L. (1979) Reproduction of heterosporous arborescent lycopods in the Mississippian–Pennsylvanian of Euramerica. *Rev. Palaeobot. Palynol.* **27** 239–289.

Phillips, T. L. (1980) Stratigraphic and geographic occurrences of permineralized coal-swamp plants: Upper Carboniferous of North America and Europe. In: Dilcher, D. L. & Taylor, T. N. (ed.), *Biostratigraphy of fossil plants*. Dowden, Stroudsburg, PA, pp. 25-92.

Phillips, T. L. (1981) Stratigraphic occurrences and vegetational patterns of Pennsylvanian pteridosperms in Euramerican coal swamps. *Rev. Palaeobot. Palynol.* **32** 5–26.

Phillips, T. L. & DiMichele, W. A. (1981) Paleoecology of Middle Pennsylvanian age coal swamps in southern Illinois: Herrin Coal Member at Sahara Mine No. 6. In: Niklas, K. J. (ed.), *Paleobotany, paleoecology, and evolution*. Praeger, New York, pp. 231–284.

Phillips, T. L. & DiMichele, W. A. (in press) Comparative ecology and life-history biology of arborescent lycopods in Late Carboniferous swamps of Euramerica. *Ann. Mo. Bot. Gard.*

Phillips, T. L. & Peppers, R. A. (1984) Changing patterns of Pennsylvanian coal swamp vegetation and implications of climatic change on coal occurrence. *Int. J. Coal Geol.* **3** 205–255.

Phillips, T. L., Kunz, A. B. & Mickish, D. J. (1977) Paleobotany of permineralized peat (coal balls) from the Herrin (No. 6) Coal Member of the Illinois Basin. In: Given, P. N. & Cohen, A. D. (eds.), *Interdisciplinary studies of peat and coal origins*. Geol. Soc. Amer. Microf. Pub. 7, pp. 18–49.

Phillips, T. L., Peppers, R. A. & DiMichele, W. A. (1985) Stratigraphic and interregional changes in Pennsylvanian coal-swamp vegetation: environmental inferences. *Int. J. Coal. Geol.* **5** 43–109.

Pianka, E. R. (1970) On *r*- and *K*-selection. *Amer. Nat.* **104** 592–597.

Pigg, K. B. & Rothwell, G. W. (1983) *Chaloneria* gen. nov.; heterosporous lycophytes from the Pennsylvanian of North America. *Bot. Gaz.* **144** 132–147.

Pimm, S. L. (1984) The complexity of stability of ecosystems. *Nature* **307** 321–326.

Piperno, D. R. (1988) *Phytolith analysis: an archaeological and geological perspective*. Academic Press, San Diego, California.

Potonié, R. (1899) *Eine Landschaft der Steinkohlen-zeit*. Borntraeger, Leipzig.

Pratt, L. M., Phillips, T. L. & Dennison, J. M. (1978) Evidence of non-vascular and plants from the Early Silurian (Llandoverian) of Virginia, U.S.A. *Rev. Palaeobot. Palynol.* **25** 121–149.

Pryor, J. S. (1988) Sampling methods for quantitative analysis of coal-balls. *Palaeogeog. Palaeoclim. Palaeoecol.* **63** 313–326.

Raymond, A. (1985) Floral diversity, phytogeography and climatic amelioration during the early Carboniferous (Dinantian). *Paleobiology* **11** 293-309.

Raymond, A., Parker, W. C. & Parrish, J. T. (1985) Phytogeography and paleoclimate of the Early Carboniferous. In: Tiffney, B. H. (ed.), *Geologic factors and the evolution of plants* Yale University Press, New Haven, pp. 169–222.

Raymond, A., Kelley, P. H. & Lutken, C. B. (1989) Polar glaciers and life at the equator: the history of Dinantian and Namurian (Carboniferous) climate. *Geology* **17** 408–411.

Raven, J. A. (1985) Comparative physiology of arthropod and plant land adaptation. *Phil. Trans. R. Soc. London B* **311** 273–288.

Read, C. B. & Mamay, S. H. (1964) *Upper Palaeozoic floral zones and floral provinces of the United States.* U. S. Geol. Surv. Prof. Pap. 434K.

Remy, W. (1975) The floral change at the Carboniferous–Permian boundary in Europe and North America. In: Barlow, J. A. (ed.), *The age of the Dunkard*. West Virginia Geol. Econ. Surv, pp. 305–352.

Remy, W. & Hass, H. A. (1991) Gametophyt aus dem Chert von Rhynie. *Argum. Palaeobot.* **8** 1–27, 29–45, 69–117.

Remy, W. & Remy, R. (1980) *Lyonophyton rhyniensis*, non. gen. et nov. spec., ein Gametophyt aus dem Chert von Rhynie (Unterdevon, Schottland). *Argum. Palaeobot.* **6** 37–72.

Remy, W., Remy, R. Leisman, G. A. & Hass, H. A. (1980) Der nachweis von *Callipteris flabellifera* (Weiss 1879) Zeiller 1898 in Kansas, U.S.A. *Argum. Palaeobot.* **6** 1–36.

Retallack, G. J. (1981) Fossil soils: indicators of ancient terrestrial environments. In: Niklas, K. J. (ed.), *Paleobotany, paleoecology and evolution, Vol. 1*. Praeger, New York, pp. 55–102.

Retallack, G. J. (1985) Fossil soils as grounds for interpreting the advent of large plants on land. *Phil. Trans. R. Soc. London B* **309** 105–142.

Retallack, G. J. (1986) The fossil record of soils. In: V. P. Wright (ed.), *Paleosols: their recognition and interpretation*. Blackwell, Oxford, pp. 1–57.

Rex, G. M. (1985) A laboratory flume tank investigation of the formation of fossil stem fills. *Sedimentology* **32** 245–255.

Rex, G. M. (1986a) Experimental modelling as an aid to interpreting the original three-dimensional structures of compressions. In: Spicer, R.A. & Thomas, B. A. (eds.), *Systematic and taxonomic approaches in palaeobotany. Syst. Ass. Spec. Vol. 31*. University Press, Oxford, pp. 17–38.

Rex, G. M. (1986b) The preservation and palaeoecology of the Lower Carboniferous plant deposits at Esnost, near Autun, France. *Geobios* **19** 773–800.

Rex, G. M. & Chaloner, W. G. (1983) The experimental formation of plant compression fossils. *Palaeontology* **26** 231–252.

Rex, G. M. & Scott, A. C. (1987) The sedimentology, palaeoecology and preservation of the Lower Carboniferous plant deposits at Pettycur, Fife, Scotland. *Geol. Mag.* **124** 43–66.

Richardson, J. B. (1985) Lower Palaeozoic sporomorphs: their stratigraphical distribution and possible affinities. *Phil. Trans. R. Soc. London B* **309** 201–203.

Robinson, J. M. (1989) Phanerozoic O_2 variation, fire, and terrestrial ecology. *Global Planet. Ch.* **75** 223–240.

Rolfe, W. D. I. (1980) Early invertebrate terrestrial faunas. In: Panchen, A. R. (ed.), *The terrestrial environment and the origin of land vertebrates. Systematics Association Spec. Vol. 15*. Academic Press, London, pp. 117–157.

Rolfe, W. D. I. (1985) Early terrestrial arthropods: a fragmentary record. *Phil. Trans. R. Soc. Lond. B* **309** 207–218.

Rolfe, W. D. I. & Brett, D. W. (1969) Fossilisation processes. In Eglington, G. & Murphy, M. J. T. (eds.), Organic geochemistry pp. 213–244.

Rothwell, G. W. (1975) The Callistophytaceae (Pteridospermopsida): 1. Vegetative structures. *Palaeontographica B* **151** 171–196.

Rothwell, G. W. (in press) *Botryopteris forensis*, an epiphyte on *Psaronius*. *Amer. J. Bot.* **78**.

Rothwell, G. W. & Mapes, G. (1988) Vegetation of a Paleozoic conifer community. *Kansas Geol. Surv. Guidebook* **6** 213–223.

Rothwell, G. W. & Scheckler, S. E. (1988) Biology of ancestral gymnosperms. In: Beck, C. B. (ed.), *Origin and evolution of gymnosperms*. Columbia University Press, New York, pp. 85–134.

Rothwell, G. W. & Warner, S. (1984) *Cordaixylon dumusum* n. sp. (Cordaitales), 1. Vegetative structures. *Bot. Gaz.* **145** 275–291.

Rowe, N. P. (1988) A herbaceous lycophyte from the Lower Carboniferous Drybrook Sandstone of the Forest of Dean, Gloucestershire. *Palaeontology* **31** 69–83.

Scheckler, S. E. (1986a) Floras of the Devonian–Mississippian transition. In: Gasataldo, R. A. & Broadhead, T. W. (eds.), *Land plants: notes for a short course*. Univ. Tennessee Dept. Geol. Sci. Stud. Geol. 15, pp. 81–96.

Scheckler, S. E. (1986b) Geology, floristics and palaeoecology of Late Devonian coal swamps from Appalachian Laurentia (U.S.A.). *Ann. Soc. géol. Belg.* **109** 209–222.

Scheckler, S. E. (1986c) Old Red Continent facies in the Late Devonian and Early Carboniferous of Appalachian North America. *Ann. Soc. Géol. Belg.* **109** 223–236.

Scheihing, M. H. & Pfefferkorn, H. W. (1980) Morphologic variation in *Alethopteris* (Pteridospermales, Carboniferous) from St. Clair, Pennsylvania. *Palaeontographica B* **172** 1–9.

Scheihing, M. H. & Pfefferkorn, H. W. (1984) The taphonomy of land plants in the Orinoco Delta: a model for the incorporation of plant parts in clastic sediments of Late Carboniferous age in North America. *Rev. Palaeobot. Palynol.* **41** 205–280.

Schopf, J. M. (1971) Notes on plant tissue preservation and mineralization in a Permian deposit of peat from Antarctica. *Amer. J. Sci.* **271** 522–543.

Schopf, J. M. (1975) Modes of fossil preservation. *Rev. Palaeobot. Palynol.* **20** 27–53.

Schopf, J. W. (1970) Precambrian microorganisms and evolutionary events prior to the origin of vascular plants. *Biol. Rev.* **45** 319–352.

Schopf, J. W. (1975) Precambrian palaeobiology: problems and perspectives. *Ann. Rev. Earth Planet. Sci.* **3** 213–249.

Scotese, C. R. (1986) *Phanerozoic reconstructions: a new look at the assembly of Asia.* Univ. Texas Inst. Geophys. Tech. Rep. 66.

Scott, A. C. (1977) A review of the ecology of Upper Carboniferous plant assemblages, with new data from Strathclyde. *Palaeontology* **20** 447–473.

Scott, A. C. (1978) Sedimentological and ecological control of Westphalian B plant assemblages from West Yorkshire. *Proc. Yorks. Geol. Soc.* **44** 461-508.

Scott, A.C. (1979) The ecology of Coal Measure floras from northern Britain. *Proc. Geol. Ass.* **90** 97–116.

Scott, A. C. (1980) The ecology of some Upper Palaeozoic floras. In: Panchen, A. (ed.), *The terrestrial environment and origin of land vertebrates. Syst. Ass. Spec. Vol. 15*. Academic Press, London, pp. 87–115.

Scott, A. C. (1984) The early history of life on land. *J. Biol. Educ.* **18** 207–219.

Scott, A. C. (1989) Observations on the nature and origin of fusain. *Int. J. Coal Geol.* **12** 443–475.

Scott, A. C. (1990) Preservation, evolution, and extinction of plants in Lower Carboniferous volcanic sequences in Scotland. In: Lockley, M. G. & Rice, A. (eds.), *Volcanism and fossil biotas*. Geol. Soc. Amer. Spec. Pap. 244, pp. 25–38.

Scott, A. C. & Chaloner, W. G. (1983) The earliest fossil conifer from the Westphalian B of Yorkshire. *Proc. R. Soc. Lond. B* **220** 163–182.

Scott, A. C. & Collinson, M. E. (1983) Investigating fossil plant beds. *Geol. Teaching* **7** 114–122, **8** 12–26.

Scott, A. C. & Galtier, J. (1988) A new Lower Carboniferous flora from East Lothian, Scotland. *Proc. Geol. Ass.* **99** 141–151.

Scott, A. C. & Rex, G. M. (1985) The formation and significance of Carboniferous coal balls. *Phil. Trans. R. Soc. Lond. B* **311** 123-137.

Scott, A. C. & Rex, G. M. (1987) The accumulation and preservation of Dinantian plants from Scotland and its borders. In: Miller, J., Adams, A. E. & Wright, V. P. (eds.), *European Dinantian environments*. Wiley, New York, pp. 329–344.

Scott, A. C. & Taylor, T. N. (1983) Plant/animal interactions during the Upper Carboniferous. *Bot. Rev.* **49** 259–307.

Scott, A. C., Galtier, J. & Clayton, G. (1984) Distribution of anatomically-preserved floras in the Lower Carboniferous in western Europe. *Trans. R. Soc. Edinburgh B* **75** 311–340.

Scott, A. C., Meyer-Berthaud, B., Galtier, J., Rex, G. M., Brindley, S. A. & Clayton, G. (1986) Studies on a new Lower Carboniferous flora from Kingswood, near Pettycur, Fife. *Rev. Palaeobot. Palynol.* **48** 161–180.

Selden, P. A. & Edwards, D. (1989) Colonisation of the land. In: Allen, K. C. & Briggs, D. E. G. (eds.), *Evolution and the fossil record*. Belhaven, Chichester, pp. 122–152.

Shear, W. A., Bonamo, P. M., Grierson, J. D., Rolfe, W. D. I., Smith, E. L. & Norton, R. A. (1984) Early land animals in North America: evidence from Devonian-age arthropods from Gilboa, New York. *Science* **224** 492–494.

Shear, W. A., Palmer, J. M., Coddington, J. A. & Bonamo, P. M. (1989) A Devonian spineret: early evidence of spiders and silk use. *Science* **246** 479–481.

Shute, C. H. & Cleal, C. J. (1987) Palaeobotany in museums. *Geol. Curator* **4** 553–559.

Signor, P. W. (1990) The geologic history of diversity. *Ann. Rev. Ecol. Syst.* **21** 509–539.

Spicer, R. A. (1977) The pre-depositional formation of some leaf impressions. *Palaeontology* **20** 907–912.

Spicer, R. A. (1981) *The sorting and deposition of allochthonous plant material in a modern environment at Silwood Lake, Silwood Park, Berkshire, England.* U. S. Geol. Surv. Prof. Pap. 1143.

Spicer, R. A. (1988) Quantitative sampling of plant megafossil assemblages. In: DiMichele, W. A. & Wing, S. L. (eds.), *Methods and applications of plant paleoecology*. Paleont. Soc. Spec. Pub. 3, pp. 29–51.

Spicer, R. A. (1989a) The formation and interpretation of plant fossil assemblages. *Adv. Bot. Res.* **16** 96–191.

Spicer, R. A. (1989b) Physiological characteristics of land plants in relation to environment through time. *Trans. R. Soc. Edinburgh B* **80** 321–329.

Spicer, R. A. & Greer, A. G. (1986) Plant taphonomy in fluvial and lacustrine systems. In: Gastaldo, R. A. & Broadhead, T. W. (ed.), *Land plants: notes for a short course*. Univ. Tennessee Dept. Geol. Sci. Stud. Geol. 15, pp. 10–26.

Spicer, R. A. & Wolfe, J. A. (1987) Taphonomy of Holocene deposits in Trinity (Clair Engle) Lake, Northern California. *Paleobiology* **13** 227–245.

Stewart, W. N. (1983) *Paleobotany and the evolution of plants.* Cambridge University Press, Cambridge.

Strother, P. K. (1989) Pre-metazoan life. In: Allen, K. C. & Briggs, D. E. G. (eds.), *Evolution and the fossil record*. Belhaven, Chichester, pp. 51–72.

Stubblefield, S. P., Taylor, T. N., Miller, C. E. & Cole, G. T. (1983) Studies of Carboniferous fungi. II. The structures and organization of *Myocarpon, Sporocarpon, Dubiocarpon,* and *Coleocarpon* (Ascomycotina). *Amer. J. Bot.* **70** 1482–1498.

Tasch, P. (1957) Flora and fauna of the Rhynie Chert: a paleoecological reevaluation of published evidence. *Univ. Wichita Bull.* **32** 1–24.

Taylor, T. N. (1988) The origin of land plants: some answers, more questions. *Taxon* **37** 805–833.

Thomas, B. A. & Brack-Hanes, S. D. (1984) A new approach to family groupings in the lycophytes. *Taxon* **33** 247–255.

Thomas, B. A. & Spicer, R. A. (1987) *The evolution and palaeobiology of land plants*. Croom Helm, London.

Tiffney, B. H. (ed.) (1989) *Phytodebris: notes for a short course*. Paleobot. Sect. Bot. Soc. Amer.

Tiffney, B. H. & Niklas, K. J. (1985) Clonal growth in land plants: a paleobotanical perspective. In: Jackson, J. B. C., Buss, L. W. & Cook, R. E. (eds.), *Population biology and evolution of clonal organisms*. Yale University Press, New Haven; pp. 35–66.

Traverse, A. T. (1988) *Paleopalynology*. Unwin & Hyman, London.

Trivett, M. L. C. V. & Rothwell, G. W. (1988) Modelling the growth architecture of fossil plants: a Paleozoic filicalean fern. *Evol. Tr. Pl.* **2** 25-29.

Valentine, J. W. (1980) Determinants of diversity in higher taxonomic categories. *Paleobiology* **6** 444–450.

Van Valen, L. (1973) A new evolutionary law. *Evol. Theory* **1** 1–30.

Vermeij, G. J. (1987) *Evolution and escalation*. Princeton University Press, Princeton.

Walton, J. (1935) Scottish Lower Carboniferous plants: the fossil hollow trees of Arran and their branches (*Lepidophloios wünschianus* Carruthers). *Trans. R. Soc. Edinburgh B* **58** 313–337.

Walton, J. (1949) A petrified example of *Alcicornopteris* (*A. hallei* sp. nov.) from the Lower Carboniferous of Dumbartonshire. *Ann. Bot. (NS)* **13** 445–452.

Walton, J. (1957) On *Protopitys* Göppert, with a description of a fertile specimen *Protopitys scotica* sp. nov. from the Calciferous Sandstone Series of Dumbartonshire. *Trans. R. Soc. Edinburgh B* **63** 333–340.

Wanless, H. R., Baroffio, J. R. & Trescott, P. C. (1969) Conditions of deposition of Pennsylvanian coal beds. In: Dapples, E. C. & Hopkins, M. E. (eds.), *Environments of coal deposition*. Geol. Soc. Amer. Spec. Pap. 114, pp. 105–142.

Westoll, T. S. (1977) Northern Britain. In House, M. R. *et al.* (eds.), *A correlation of the Devonian rocks of the British Isles*. pp. 66–93. Geol. Soc. Lond. Spec. Rep. 8.

Wiley, E. O. (1981) *Phylogenetic systematics*. Wiley, New York.

Wiley, E. O., Siegel-Causey, D., Brooks, D. R. & Funk, V. A. (in press) *The compleat cladist: a primer of phylogenetic procedures*. Univ. Kansas Mus. Nat. Hist., Lawrence, KS.

Willard, D. A. (1989) Source plants for Carboniferous microspores. *Amer. J. Bot.* **76** 820–827, 1429–1440.

Wing, S. L. (1984) Relation of paleovegetation to geometry and cyclicity of some fluvial carbonaceous deposits. *J. Sedim. Petrol.* **54** 52–66.

Wing, S. L., DiMichele, W. A., Mazer, S., Phillips, T. L., Spaulding, G. & Tiffney, B. H. (in press) Ecological analysis of fossil plants. In ETE Consortium (eds.), *Evolutionary paleoecology of plants and animals*. Chicago University Press, Chicago.

Winston, R. B. (1988) Paleoecology of Middle Pennsylvanian-age peat-swamp plants in Herrin Coal, Kentucky, U.S.A. *Int. J. Coal Geol.* **10** 203–238.

Winston, R. B. (1990) Implications of paleobotany of Pennsylvanian-age coal of the central Appalachian Basin for climate and coal-bed development. *Geol. Soc. Amer. Bull.* **102** 1720–1726.

Wnuk, C. (1985) The ontogeny and paleoecology of *Lepidodendron rimosum* and *Lepidodendron bretonense* trees from the Middle Pennsylvanian of the Bernice Basin (Sullivan County, Pennsylvania). *Palaeontographica* **195** 153–181.

Wnuk, C. (1989) Ontogeny and paleoecology of the Middle Pennsylvanian arborescent lycopod *Bothrodendron punctatum*, Bothrodendraceae (Western Middle Anthracite Field, Shamokin Quadrangle, Pennsylvania). *Amer. J. Bot.* **76** 966–980.

Wnuk, C. & Pfefferkorn, H. W. (1984) The life habit and paleoecology of Middle Pennsylvanian medullosan pteridosperms based on an *in situ* assemblage from the Bernice Basin (Sullivan County, Pennsylvania, U.S.A.). *Rev. Palaeobot. Palynol.* **41** 329–351.

Wnuk, C. & Pfefferkorn, H. W. (1987) A Pennsylvanian-age terrestrial storm deposit: using plant fossils to characterize the history and process of sediment accumulation. *J. Sedim. Petrol.* **57** 212–221.

Wolfe, J. A. & Upchurch, G. R. Jr. (1987) Leaf assemblages across the Cretaceous–Tertiary boundary in the Raton Basin, New Mexico and Colorado. *Proc. Natl. Acad. Sci. USA* **84** 5096–5100.

Wright, V. P. (1985) The precursor environment for vascular plant colonization. *Phil Trans. R. Soc. London B* **309** 143–145.

Ziegler, A. M. (1990) Phytogeographic patterns and continental configurations during the Permian Period. In: McKerrow, W. S. & Scotese, C. R. (eds.), *Palaeozoic palaeogeography and biogeography. Geol. Soc. Lond. Mem.* **12** 363–379.

Ziegler, A. M., Bambach, R. K., Parrish, J. T., Barrett, S. F., Gierlowski, E. H., Parker, W. C., Raymond, A. & Sepkoski, J. J. Jr. (1981) Paleozoic biogeography and climatology. In: Niklas, K. J. (ed.), *Paleobotany, paleoecology and evolution*. Praeger, New York, pp. 231–266.

3

Silurian and Devonian

D. Edwards and **C. Berry**

A bad workman blames his tools, the palaeobotanist working in the Silurian and Devonian the database. And not without justification — it cannot be denied that the dearth of plant fossil occurrences prevents global generalizations on biostratigraphy and palaeophytogeography as it does the detection of evolutionary patterns.

However, the resurgence of interest in early land-plant fossils seen in the last twenty-five years has produced extra information and a far more critical approach to its analysis and, most important, to age determination. Many of the problems relating to such quality control are equally relevant to analyses higher in the Palaeozoic, but are exacerbated here by the unique evolutionary position of the plants themselves, encompassing the early stages of diversification on land particularly of the vascular plants. Most evidence comes from Laurasia, where the remains of exceedingly simple plants (rhyniophytoids), archetypal in concept, precede more complex forms (Edwards & Davies 1990). Yet in Australia, new Ludlow assemblages (Tims & Chambers 1984) contain the 'prelycophyte' *Baragwanathia*, a plant with organization not seen in the present northern hemisphere until higher in the Gedinnian (Lochkovian), with *Drepanophycus spinaeformis* (Schweitzer 1983); *Baragwanathia* itself is not known in north Laurasia (Ontario) below the Emsian (Hueber 1983). This is an excellent demonstration of the pitfalls of geographical extrapolation away from the more intensively studied Laurasia, and highlights a major theme of this chapter.

3.1 HISTORICAL PERSPECTIVE (PRIOR TO 1968)

Arber (1921) is credited with the earliest consideration of Devonian plant fossil assemblages in a stratigraphical setting, when he established two overlapping 'floras'.

(1) A lower, *Psilophyton* 'flora', with predominantly axially organized plants including the earliest land plants in the lower part, and progymnosperms

(excluding 'leafy' *Archaeopteris*) and certain early ferns *sensu lato* e.g. *Pseudosporochnus* higher up.

(2) An upper, *Archaeopteris* 'flora', with the above-mentioned progymnosperms (but this time including *Archaeopteris* high in the zone) and 'ferns', but also including various lycophytes, sphenophytes and 'ferns'. The two zones overlapped in the Middle Devonian, and so Kräusel (1937) introduced a third zone — the *Hyenia* 'flora' — for the transitional assemblages. The resulting three 'floras' thus correspond to the three major subdivisions of the Devonian. This scheme was followed by Leclercq (1940), except that she replaced *Hyenia* with *Protopteridium* (now *Rellimia*).

3.2 QUALITY CONTROLS ON THE DATABASE

To facilitate discussion on existing biostratigraphical schemes and palaeophytogeographical analyses, the stratigraphical ranges and geographical occurrences of selected form-genera have been plotted (Figs 3.1–3.3, 3.5–3.8). In preparing these figures an attempt has been made to look critically at a sometimes unfamiliar database; a number of limitations have been encountered.

3.2.1 Geographical coverage and location

Figs 3.9–3.16 show the Silurian and Devonian occurrences to be concentrated on the palaeocontinent Laurasia, thus reflecting the activity of American and European workers. A drawback of such maps is the lack of information on the number of localities associated with each record (but see Raymond *et al.* for a more useful approach). Thus, for example, the Wenlock record in Ireland represents a small collection in a ploughed field possibly from one horizon (Edwards *et al.* 1983). In contrast, the Gaspé record represents a collection from a number of distinct horizons over a considerable distance of shoreline (Gensel & Andrews 1984), while those from the Siegenian–Emsian of Belgium and Germany mark several localities over an extended area (Stockmans 1940; Schweitzer 1983).

In the case of some less well documented or old records (e.g. Poland — Gothan & Zimmermann 1937) which lie very close to what are now regarded as plate boundaries and which have been involved in continental collisions after the plants were deposited, there is the added problem of deciding on which plates to plot the localities. In the Polish example, and other similar examples where the present authors have no 'hands-on' knowledge, these have simply been plotted as accurately as possible on the base maps, and no account has been taken of possible compression and overthrusting which may have occurred. The comments of experts with the appropriate local geological knowledge are awaited.

Sources of optimism for more global coverage are recent records from China (see references in Hao 1989; Li & Edwards in press), Greenland (Larsen *et al.* 1986), and South America (Cuerda *et al.* 1987). Collections made from a number of coeval localities in a restricted geographical area allow elimination or estimation of facies bias (see e.g. Edwards 1990a).

In addition to the points plotted, Middle Devonian plant fossils are known from northwestern China (Dou & Sun 1983, Lee & Tsai, 1978). However, the exact positions of the localities could not be determined, and so they have been omitted.

3.2.2 Stratigraphical coverage and correlation

Sensible biostratigraphical analysis requires as continuous a stratigraphical record as possible from a number of geographical areas, preferably in close proximity in the first instance. This has not been achieved for Devonian plant fossils, although there are some excellent 'monographic' works for some areas (e.g. Schweitzer — Lower Devonian of the Rhineland; Gerrienne — Lower Devonian of the Ardenne; Bonamo, Banks *et al.* — Givetian–Famennian of the Catskill wedge New York State). Even in the latter, Banks (1980) highlighted problems of correlation in continental sediments where actual dating is difficult and the deposits themselves are mere lenses. A reconsideration of the positioning of assemblages in the light of revision of local stratigraphy (see e.g. Woodrow *et al.* 1988) may produce greater understanding. Relating to Europe, Edwards & Davies (1990), compiled data from the Silurian and Lower Devonian in palaeogeographically close occurrences in Britain, Belgium and Germany, but found little evidence for stratigraphical overlap and the coeval duplication of records. The same range charts are characterized by high numbers of single occurrences, again emphasizing monographic treatment (e.g. Silurian records in the Welsh Borders) and/or exceptional preservation (e.g. Rhynie Chert). In other examples some endemics may be recorded under different names when preserved differently. Thus, for example, the collection of axes called *Karagandella* in Kazakhstan (Yurina 1965) may be congeneric with petrified progymnosperms from elsewhere.

Exclusion of endemics thus greatly reduces the number of form-genera in our stratigraphical range chart. A further limitation in our own and others' range charts is their failure to convey information on the presence or absence of the taxa in the interval between the first and last records.

As is the case with accurate stratigraphical documentation of occurrences, there has been major improvement since Banks (1980) cited problems of correlation as a reason for the tentative nature of his zonal scheme. Correlation of horizons in continental sediments in Laurasia in terms of standard international sequences has been greatly facilitated by reference to comprehensive miospore-based sequences developed by a number of palynologists (e.g. Streel, Steemans, Richardson, McGregor) throughout the Silurian and Devonian. Most important is the recent zonation of Richardson & McGregor (1986) and for the topmost Devonian, the intensity of activity related to the positioning of the Devonian–Carboniferous boundary (e.g. Becker *et al.* 1974). Problems still exist. In the Lower Devonian, good correlation has been achieved between southern Britain and the Ardenne, and indirectly via Podolia with deep marine sediments of the international stratotypes in Czechoslovakia (Richardson *et al.* 1981). However, precise correlation close to the Gedinnian–Siegenian boundary between Britain and Czechoslovakia, and hence the Lockkovian–Pragian boundary, a level of enormous importance in the early diversification of vascular plant fossils, has not yet been achieved. A further problem relates to detailed palynological documentation of the formations containing plant fossil assemblages in the Middle and Upper Devonian of New York State, and hence lack of precision in correlation with records elsewhere in Laurasia. More successful application of palynology in the dating of plant fossils has been demonstrated by Lessuise & Fairon-Demaret (1980) in their rigorous analysis of Eifelian–Givetian records from the Ardenne and Rhineland.

Independent faunal evidence for dating is rare in continental sediments. An exception is found in the Lower Devonian of the Welsh Borders (although not in the Brecon Beacons) where abundant fish remains are interbedded with plants (Ball & Dineley 1961).

In contrast, as already mentioned, marine Silurian assemblages are usually more confidently dated using graptolites (Richardson & Edwards 1989), as are some Lower Devonian localities in north America (Churkin *et al*. 1969).

The age of plant fossil assemblages outside Laurasia are clearly of fundamental importance to the development of a global stratigraphy and thus recent progress is worthy of greater emphasis than is perhaps expected in a review of this kind.

3.2.2.1 Gondwana

The recognition of *Baragwanathia* assemblages as Silurian from graptolitic evidence is already well-known, although the fact that these assemblages are relatively recently discovered merits repeating. This was not a case of the re-interpretation of the identity and hence ages of graptolites which were originally thought to be Silurian (Lang & Cookson 1935) and then Devonian (Jaeger 1962). In Victoria, plant fossil assemblages associated with graptolites occur in upper Silurian to Emsian sediments, the lowest and highest being separated by several thousand metres of sediment. Palynological data for the area is still lacking. Higher in the Devonian, Australian assemblages so far described, although less diverse than those in Laurasia, are noteworthy only for their similarities, while in the Upper Devonian they contain abundant *Leptophloeum*.

A second, frequently cited, anomaly comes from Libya, where dating of presumed Silurian and Lower Devonian assemblages is controversial and poorly constrained (e.g. Lejal-Nicol 1975), and the identifications themselves suspect. An exception to the former is the very poorly preserved *Cooksonia* found in a borehole and dated by graptolites (Daber 1971). Despite major progress achieved on Silurian and Devonian palynostratigraphy in the region (Richardson 1988; Streel *et al*. 1990b) these successions are often considerable distances from plant-bearing horizons, on which there is still no direct evidence for age. Streel *et al*. (1990b) questioned the existing lithostratigraphic correlation on the grounds of diachronism. The stratigraphically lowest assemblages from the Acacus Formation, thought possibly as low as Wenlock (Klitzsch *et al*. 1973), contain very fragmentary fossils broadly described as 'psilophytes', with some assigned (perhaps over-optimistically — see Edwards 1990a) to well-established form-genera such as *Steganotheca* and *Dawsonites*. Streel *et al*. concluded that these fragments could not have originated from below the Pragian and could even be from the Emsian, on the basis of axis and (presumed) sporangium dimensions, and on branching pattern. Such inferences may well be correct (and justified in the context of Laurasia), but are less well founded when comparison is made with dimensions of plants in the Silurian assemblages of

Australia (Tims & Chambers 1984). The presumed younger lycophytes from the Tadrart Formation appear to be short lengths of stems with persistent leaves or showing differing amounts of decortication. There are undoubtedly herbaceous lycophytes broadly similar to those described elsewhere in the Middle and Upper Devonian, but their attribution to form-genera such as *Protolepidodendon* and *Archaeosigillaria* is premature (see discussion in Edwards & Benedetto 1985). A detailed and critical analysis of the lycophytes previously considered as Lower Devonian (Lejal-Nicol & Moreau-Benoit 1979) further highlights the problems of identifying fragmentary and often decorticated herbaceous and arborescent lycophytes. It is of particular interest to this book that the Belgian workers dated the Tadrart and Emi Magri formations as Middle Devonian on the basis of the plants themselves. Information on assemblages from elsewhere in Gondwana has recently been summarized by Edwards (1990b).

3.2.2.2 Soviet Union

Occurrences are numerous and stratigraphically widespread, but limited mainly to three geographical areas, viz, the Russian platform (Podolia–Urals–Timan), central Kazakhstan and the Altai–Saian region of south-west Siberia (for an excellent map, see Yurina, 1988, p. 95). However, interpretation of the extensive database both systematically and biostratigraphically is difficult for western workers, and assessments based on mere scrutiny of often poorly illustrated species lists are unreliable. This is in part due to the fact that most Soviet palaeobotanists are geologists employed as biostratigraphers using plants themselves in dating and correlations. Their identifications, often based on sterile, poorly preserved fragments which rarely are, or indeed can be, subjected to rigorous analysis, are sometimes over-optimistic using familiar Laurentian taxa. But deficiencies are not confined to Soviet workers. Interpretations here can be taken in another sense — that of translation. The prohibitive costs of professional translation of long papers in Russian prevent adequate appreciation of papers, often with a strong stratigraphical bias, by workers such as Stepanov (1975), Senkevich (e.g. Kaplan & Senkevich 1977) and Yurina (1988).

3.2.2.3 China

The renaissance of Devonian palaeobotany in China began with the publication of 'Devonian floras of China' by Lee & Tsai (1978) for the Bristol P.A.D.S. meeting. This summary placed plant fossils in a stratigraphical framework, and showed a substantial Laurasian influence. More recent fieldwork, for example in Yunnan Province, has been attended by a far greater awareness of the need for detailed and preferably independent information on age, and has resulted in the accumulation of quite spectacular fossils. In the Lower Devonian of Yunnan, assemblages containing a high percentage of endemics (Li & Edwards in press) are accurately located stratigraphically, although correlation still lacks the precision of Laurasia (e.g. Hao 1989). Detailed palynostratigraphy has not yet been achieved. These assemblages are in continental facies and thus contrast with a graptolite-associated record in the Junggar Basin, Xinjiang Province, north-west China. Dou & Sun (1985) originally dated these facies as Gedinnian on the basis of the graptolites, and the K- and H-shaped branching typifying *Zosterophyllum*, then thought to characterize plant fossils from the base of the Devonian. In the same assemblage, they described two endemic rhyniophytoids, *Salopella xingiangensis* and *Junggaria spinosa*, together with abundant sterile axes including *Hostinella*, *Psilophytites* and a single fragment of a putative lycophyte. Further collections of graptolites from the plant-containing middle part of the Wutubulake Formation indicated that the sediments were indeed Silurian and in the *bouceki/bandaletovi* graptolite biozone towards the top of the Pridoli. A similar revision of graptolites from the Tokrau Formation in the Balkhash region of neighbouring Kazakhstan had also suggested a Pridoli age (Nikitin & Bandaletov 1986). In this case, associated land plants included *Taeniocrada* sp., a sterile leafy axis thought to resemble the lycophyte *Baragwanathia*, *Jugumella burubaenis* and *Cooksonella sphaerica*. The last is now considered synonymous with *Junggaria spinosa*, and suggests a close proximity of the plant assemblages in Silurian times. When Edwards (1990a) first indicated such similarities, the Xinjiang locality was thought to be on the Siberian plate, but more recent and complex reconstructions by Chinese workers place it on a small Kazakhstan plate, clearly part of the Kazakhstan complex. Thus doubts, based on plant fossils, on the palaeogeography of certain parts of Xinjiang have been vindicated — a good example of the use of fossil plants in palaeogeography.

Despite such progress, more records are required before the biostratigraphic value of these assemblages can be ascertained. It is possible that the Kazakhstan/Xinjiang assemblages demonstrate a second phytochorion in the topmost Silurian, but all the records for coeval Laurasia are in the early Pridoli.

Similarities with Kazakhstan were also noted by Schweitzer & Cai (1987) in a reassessment of Middle Devonian assemblages from South China. This paper, together with those by Li & Hsu (1987) and Li (1990), provides the kind of basic information needed for global analysis. In the Upper Devonian, recent reports include those on *Archaeopteris* (Cai 1981; Cai, *et al.* 1987) from South China and *Leptophloeum* and *Callixylon* in diverse assemblages from the north-west (Dou & Sun 1983, Cai & Qin 1986; Cai 1989).

3.2.3 Taphonomic and palaeoecological considerations

For the most part, the fossil record for vascular plants is dominated by those growing near water/land interfaces, thus presenting a somewhat biased picture of land vegetation as a whole. However, early in the colonization of the land by higher plants with exclusively homosporous reproduction, the record might well be more representative.

The wide variety of fluvial and lacustrine sediments comprising the Old Red Sandstone would undoubtedly have promoted some sorting of fragmentary fossils during deposition (e.g. Edwards & Fanning 1985, Edwards 1990a) and should be taken into account in evaluating the palaeophytogeographical significance of single occurrences, although Raymond *et al.* (1985) did not consider that attached sporangia influenced sorting of Lower Devonian adpression fossils.

Plant fossil assemblages in marine strata, with implication of sometimes considerable transport, are particularly important in the Silurian. Here, *all* occurrences are marine, with diversity increasing proportionally in younger and progressively shallower sediments. Indeed, Pridoli examples in the Welsh Borderland are found in coastal, intertidal sediments.

This pattern thus appears to reflect the initial radiation of rhyniophytoids (Edwards & Davies 1990) but as yet it is impossible to evaluate to what extent transport influences the record. A similar problem exists with low-diversity assemblages from north-east and northern Laurasia, both dated by graptolites. Raymond *et al.* (1985) and Raymond (1987) attached palaeophytogeographical significance to the Bathurst Island assemblage (Hueber 1971) believing it to represent a subunit of Laurentia with links with Siberia (presumably due to the common occurrence of *Rebuchia*). However, in both Canada and Alaska (Churkin *et al.* 1969) low diversity could equally result from the environment of deposition.

A further feature of note, but one which as yet does not appear to have distorted the record, is the tendency for certain horizons, and hence individual bedding planes, to be dominated by a single taxon. This reflects mode of growth in monospecific stands and proximity of growth to depositional environments. The Lower Devonian Gaspé record is a succession of shoreline outcrops, each dominated by a single species, only rarely a mixture (Gensel & Andrews 1984). Domination by a single species, *Rhacophyton ceratangium*, is also recorded in the uppermost Devonian, but here it is in coals (Scheckler 1986a) and is thus the earliest record of convincing swamp vegetation.

The influence of local variation in vegetation on the composition of assemblages again requires detailed analysis of a number of occurrences in a small area and has repercussions for both palaeophytogeographical and biostratigraphical analyses. Palaeoecological studies are rare in the Devonian; two important recent examples include Scheckler's work on Upper Devonian coal swamp deposits from Appalachian Laurasia (1986a) and Schweitzer's reconstruction of coastal vegetation from the Lower Devonian of the Rhineland (1983).

3.2.4 Quality of preservation/identification

The nature of the vascular plant fossil record is unique in the Silurian and Devonian, in that the majority of the fossils are of an axial nature and can be identified with certainty only when fertile, or show anatomical detail. Examples of the latter include the cosmopolitan and long-ranging taxa *Sawdonia ornata* and *Drepanophycus spinaeformis*. Both possess very distinctive cuticular/epidermal features and these should be demonstrated before positive identification is made. Indeed, in this account, the use of the form-genus *Drepanophycus* is confined to the type species, in that similar reproductive characteristics have not been demonstrated in other species. Considering *Sawdonia*, broadly similar spinous axes characterize *Psilophyton goldschmidtii/burnotense*, which are sometimes preserved with marginal features, and have frequently been recorded in the Soviet Union. On the discovery of attached lateral sporangia in Siberian fossils, Zakharova (1981) transferred them to *Margophyton goldschmidtii*, but Schweitzer (in press) has recently shown European examples to be trimerophytes. In such instances, it would be helpful if a symbol could be appended to species lists where records are based only on sterile remains. This would be particularly appropriate for records of *Zosterophyllum* based only on Hand K-shaped configurations in smooth axes, and also for *Taeniocrada*. The latter form-genus clearly embraces taxa of diverse affinity (e.g. *Stockmansella (Taeniocrada) langii*), united in the presence of presumably strap-shaped axes, each with a very narrow central-conducting strand. Unfortunately, the type species, *T. lesquereuxi*, is sterile and thus has little biostratigraphical use. In the range chart reference is made only to *T. decheniana*, but it is felt that this should also probably be transferred to a new form-genus.

A further problematic and commonly cited widespread Devonian form-genus is *Protolepidodendron*. The problem stems from two sources:

(1) the very complete description of the lycophyte *Leclercqia complexa* from the Middle Devonian of New York State (Banks *et al.* 1972); and,
(2) the transfer of *P. wahnbachense* to the new form-genus *Estinnophyton* (Fairon-Demaret 1978). Following the first, it became obvious that many of the specimens assigned to *P. scharyanum* did indeed have leaves with five points, and hence belonged to *Leclercqia* (e.g. material from Australia — Fairon-Demaret 1974). However, development of the type specimens themselves, in which the illustrated leaves are simply forked, has not yet been possible and according to Dr Fairon-Demaret, as the matrix is coarse, may not be conclusive. It is thus tempting but premature to assume that *Protolepidodendron scharyanum* and *Leclercqia complexa* are synonymous. This leaves the predicament of how to interpret the records of species of *Protolepidodendron* outside Laurasia. A partial solution has come from China, where Li (1990) has recently redescribed Chinese Middle Devonian *P. scharyanum* as the new form-genus *Minarodendron*.

In the range chart, therefore, *Protolepidodendron* is restricted to *P. scharyanum* from the type locality, it being retained because it is a particularly familiar name to palaeobotanists.

Identification based on inadequately known sterile and exceedingly fragmentary fertile remains were described as optimistic by Edwards (1990a), who cited examples from Libya and the Sahara as particularly misleading. Similarly Raymond (1987), in an attempt to eliminate possible identification difficulties (regional taxonomic bias), used the distribution of morphological features in a palaeophytogeographical analysis.

In attempting to demonstrate a biostratigraphical use for sterile axes, Chaloner & Sheerin (1979) showed that average axis diameter increases throughout the Devonian and so, taking sorting and overall branching patterns into account, it may well be possible to impose a maximum age on sediments enclosing the axial plant fossils, at least in Laurasia.

Many taxa described from Middle and Upper Devonian sediments in the United States are based on petrifactions and, although in a few cases are accompanied by reconstructions of gross morphology (e.g. *Triloboxylon ashlandicum* — Scheckler 1975), the majority cannot be related to adpressions found elsewhere. A further problem arises in assignment of petrified axes to taxa known primarily as adpressions, especially as there are examples of convergence in vascular anatomy. The New York State Cairo assemblage contains two pyritized axes named as *Rhacophyton ceratangium* by Matten (1974). As this record markedly extends the range of a form-genus otherwise found in middle and upper Famennian and basal Carboniferous sediments, and indeed is the nominal taxa for Banks Zone VII, it is assumed that Banks omitted it from his deliberations in the absence of diagnostic morphological evidence.

Important exceptions to this axial fossil record are seen in the lycophytes. In the case of herbaceous forms with persistent leaves, form-genera can be relatively easily identified on leaf morphology alone, although individual specimens may need development to reveal detail (Bonamo, *et al.* 1988). More difficult to identify are plants whose stem surface characteristics are reminiscent of those in Carboniferous lepidophytes, but show various levels of decortication, such that diagnostic features are obscured. Examples here include abundant remains from Libya (Lejal-Nicol & Moreau-Benoit 1979, and references therein), Kazakhstan (see summary in Senkevich 1982) and China (Dou & Sun 1983).

Finally, in the earliest stages of higher plant diversification, recent research has shown that plants with identical morphology can be distinguished at the anatomical level (Fanning *et al.* 1988). This clearly has implications in studies of diversity and floristic analysis but, in that all representatives to date occur in sediments of approximately the same age, it is not a problem in biostratigraphy.

3.3 TOWARDS A TENTATIVE SYNTHESIS

In order to appreciate the existing biostratigraphical and palaeophytogeographical analyses reviewed below, the reader needs some immediate information on the new data. The range charts in Figs 3.1–3.3 thus represents the relevant analysis of the

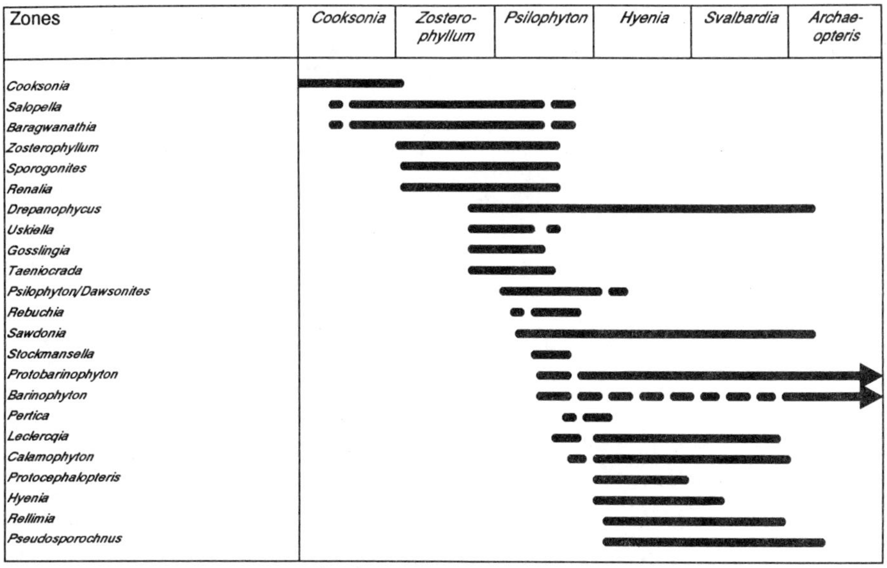

Fig. 3.1 — Stratigraphical ranges of key Silurian and Devonian taxa. See text for sources of data.

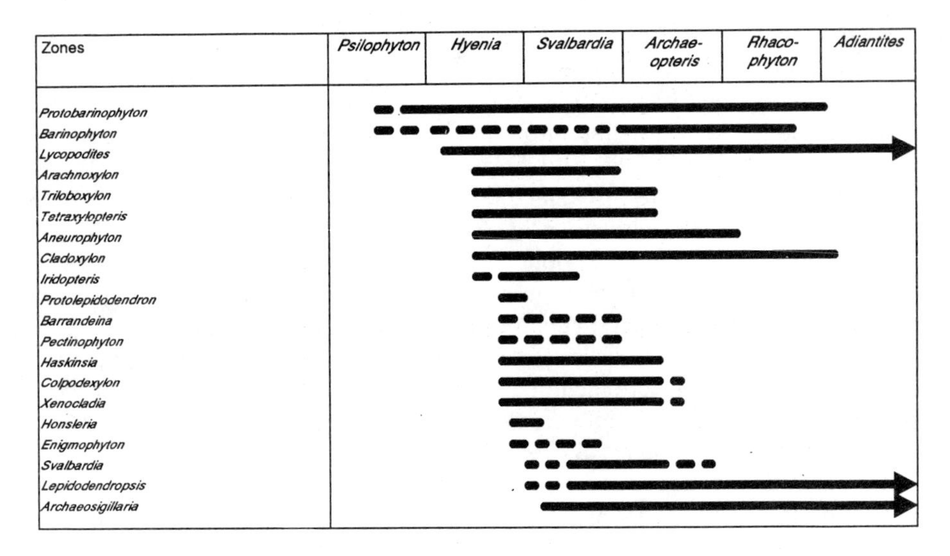

Fig. 3.2 — Stratigraphical ranges of key Devonian taxa. See text for sources of data.

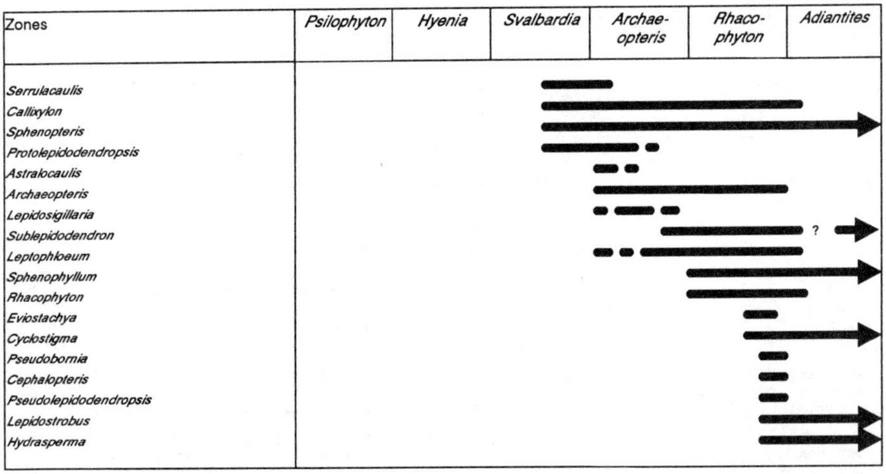

Fig. 3.3 — Stratigraphical ranges of key Devonian taxa. See text for sources of data.

current database taken from a major literature survey. For the Silurian and Lower Devonian this has already been partly summarized in Edwards (1990a) and Edwards & Davies (1990). Many of the sources for data later in the Devonian, e.g. taxonomic reassessments, new localities, are referred to elsewhere in this text. In selecting taxa the authors have concentrated, as Banks (1980) did, on Laurasia. Endemics have been omitted. For reasons of space, form-genera have been used and not species, but alongside this pragmatism was the feeling that generic identifications are more reliable, particularly where the authors have no intimate acquaintance with the taxa concerned. Here it may be of interest to compare the numbers of species analysed in the Silurian and Lower Devonian by Edwards & Davies (1990). The form-genera cited in the present range chart are similar to those in Chaloner & Sheerin (1979) with surprisingly little modification after ten years of research, while the duration of Banks' zones and the composition of their assemblages remain relatively unchanged. Most frustrating in trying to achieve precision in stratigraphical documentation are those taxa cited as, for example, Lower Devonian or Famennian, without further qualification. These instances are characterized by a broken line. A questionmark indicates where identity rather then age is questioned.

3.4 PLANT FOSSILS AND BIOSTRATIGRAPHY

In 1980, Banks brought together the newly acquired data in an attempt to demonstrate the potential of plant fossils in biostratigraphy, but at the same time emphasizing reasons for the provisional nature of his zonation scheme. This comprised seven biozones based on generic assemblages, with the nominal taxon (thought representative or typical) usually marking the base of each biozone, and then extending (normally) beyond its upper limit. The biozonation was essentially based on Laurasia, but no type sections were designated. Inadequacies of records outside Laurasia continue to be a frustration, especially as some of the new information hints at the development of provincialism.

Banks combined the more conventional biostratigraphical approach with one best described as an 'alternative biostratigraphy' relating to morphological and anatomical characteristics displayed by the plant fossils (see also Banks 1981). Chaloner & Sheerin (1979) had already documented the first appearances of such characters, or 'bioevents', in the Devonian, emphasizing the role of evolutionary innovation in establishing a maximum age for sediments. Thus, for example, sediments containing ovules are probably not older than Famennian, and those containing secondary wood, Eifelian. This perhaps denies a role for the fossil record in establishing centres of origin, arguably not unlikely in the light of the inadequate nature of the plant fossil record itself.

3.4.1 Assemblage Biozone I — *Cooksonia* Biozone

Chronostratigraphic range — Wenlock (*ludensis* graptolite zone) to top of Pridoli.

Since 1980, *Cooksonia* has been recorded from the Wenlock of Ireland (Edwards *et al.* 1983), thus extending the zone into the Wenlock, while the discovery of *Ambitisporites* in Llandovery sediments leads to the prediction that *Cooksonia* will eventually be found even lower. Again, the zone is essentially based on Laurasia, the most notable geographical extension in the palaeocontinent being the discovery of *Salopella* in northern Greenland (Larsen *et al.* 1986). The Libyan borehole record in graptolitic facies still holds (Daber 1971), but the dating of anomalous Libyan assemblages (cited as Silurian by Lejal-Nicol (1975)), has recently been revised to at least Pragian (Streel *et al.* 1990). The zone thus remains applicable to Laurentia and north Gondwana (Bohemia and Libya) but cannot be applied to Australia, where Ludlow assemblages contain much larger vascular plants and, in the case of *Baragwanathia*, of more complex organization and hence affinity.

Considering morphological characteristics, additions to Banks' list include spines on axes and more remarkably on sporangia (Fanning *et al.* 1990, Fanning *et al.* in press), and the demonstration of stomata (Jeram *et al.* 1990). While vascular tissues have been demonstrated in Ludlow sterile axes from Wales and their presence inferred in Australian *Baragwanathia*, there remain no unequivocal fertile rhyniophytes in the Silurian, and *Cooksonia* should still be described as rhyniophytoid.

3.4.2 Assemblage Biozone II — *Zosterophyllum* Biozone

Chronostratigraphic range — Gedinnian to lower Siegenian.

Zosterophyllum myretonianum is found low in the Gedinnian of Scotland and remains the stratigraphically lowest record, although there are reports, but not yet descriptions, of zosterophylls in Australia (Douglas & Lejal-Nicol 1981, Garratt & Rickards, 1984). *Zosterophyllum* from the Pridoli of Xinjiang is based on vegetative remains and should be discounted. A recent review (Edwards 1990a) indicates diversity of rhyniophytoids associated with early *Zosterophyllum*, although rarely in the same assemblages, perhaps as a result of sorting or local geographical variation. Relevant to Banks' zonation scheme is the discovery in South Wales of a diverse assemblage of zosterophylls near the top of the Gedinnian (Edwards in press; Edwards *et al.* 1989), and the occurrence of *Drepanophycus spinaeformis* in sediments of similar age in Germany (Schweitzer 1983). Thus many of the taxa, and hence characteristics, that Banks considered to have appeared in the mid-Siegenian already existed in the upper Gedinnian, although the latter consist of adpressions with no supporting anatomical data. Gedinnian geographical occurrences (Edwards 1990a,b) are again concentrated on Laurasia, a notable exception being that from the Kuznetsk basin, Siberia (Stepanov 1975) where *none* of the European taxa are unequivocally recorded.

3.4.3 Assemblage Biozone III — *Psilophyton* Biozone

Chronostratigraphic range — lower Siegenian to top of Emsian.

The advent of some of the diversity that Banks thought characteristic of his Biozone III is now known to have occurred earlier (see above). Further, because the Brecon Beacons assemblages have been re-dated on spores as being closer to the base of the Siegenian, this changes the level of introduction of the trimerophytes as well as the advent of anatomical characteristics mentioned by Banks. Trusses of sporangia of *Dawsonites* type, attached to smooth axes, occur in the Brecon Beacons Quarry and slightly younger Llanover Quarry (Edwards & Davies 1990). More fragmentary *Dawsonites* recorded from the Siegenian of Australia is generically less convincing.

Banks himself suggested a possible further subdivision at the Siegenian–Emsian boundary or perhaps even higher, in the Emsian. The stratigraphical range chart for

Laurentia appears to indicate an increase in diversity (i.e. in numbers of taxa) in the low Emsian (although this is partly due to imprecise dating, where occurrences are cited as Emsian), but innovation, with plants of progressively more advanced organization, is evident throughout the Emsian.

3.4.4 Assemblage Biozone IV — *Hyenia* Biozone

Chronostratigraphic range — topmost Emsian to top of Eifelian.

Fairon-Demaret (1986) commented on the usefulness of Banks' zonation scheme, and its facility of application except for assemblages near boundaries 'because of unavoidable overlap of critical form-genera resulting either from their natural longevity or from looseness of their known stratigraphical records'. This is certainly true for the *Hyenia* Biozone. Thus, for example in the United States, a number of Emsian form-genera extend into the Eifelian, while in Siberia there is some controversy on the age of highest Emsian assemblages. Identification of *Hyenia* itself is sometimes controversial, and its morphology and anatomy are incompletely known when compared with *Calamophyton*. Indeed the latter may well be a better choice for the nominal form-genus (M. Fairon-Demaret, pers. comm.). The latter is recorded from the base of the Eifelian, although unfamiliar species may extend down into the Emsian in the Soviet Union (such occurrences required careful scrutiny outside Laurasia). *Calamophyton*, which Banks recommends, together with *Hyenia*, as marking the base of the zone, is now recorded from uppermost Emsian of Belgium (Lessuise & Fairon-Demaret 1980). Yurina (1988) thus suggests a downward extension of Zone IV to accommodate horizons in western Europe with *C. primaevum* and *H. elegans*. The zone itself sees the proliferation and diversification of the progymnosperms, with attendant problems of identification of sterile axial systems and petrifactions.

3.4.5 Assemblage Biozone V — *Svalbardia* Biozone

Chronostratigraphic range — much of the Givetian.

Banks (1980) himself pinpointed the problems likely to be encountered in the application of this zone, and indeed the succeeding *Archaeopteris* biozone, in that it contains many long-ranging form-genera, admittedly more geographically abundant, such as *Colpodexylon*, *Aneurophyton*, *Rellimia*, and *Pseudosporochnus*. However, he chose to distinguish it by the advent of the more advanced group of progymnosperms, the Archaeopteridales, just above the base of the Eifelian–Givetian boundary, its lower limits marked by the appearance of the petrifaction *Actinoxylon* of archaeopterid affinity. Although Banks felt that detailed searching would reveal the existence of this form-genus outside New York State, his expectation has not yet been realized (perhaps on preservational grounds) and the zone is perhaps better characterized by the appearance of the more widespread *Svalbardia*. However, even the recognition of this form-genus is not without problems, (see, for example, discussion on *Archaeopteris* fossils in Scheckler 1978 and Matten 1981) and indeed Banks' own paper. The most biologically significant advance during this time interval was the emergence of the arborescent habit both in certain progymnosperms and in the lycophytes. The latter display superficial features of lepidodendrids (e.g. the diamond-shaped leaf bases in *Lepidodendropsis*), although there is still some debate on the exact relationship of these structures to the leaf cushions of Carboniferous lepidodendrids, and there are problems of identification, particularly in the Soviet Union and China.

3.4.6 Assemblage Biozone VI — *Archaeopteris* Biozone

Chronostratigraphic range — Frasnian to middle Famennian (Fa 1/2 boundary).

Banks marked the base of this zone by the appearance of *Archaeopteris* species with planated, webbed leaves in assemblages containing long-ranging Givetian taxa. The zone was characterized by the abundance and wide distribution of *Archaeopteris* foliage and its wood (at least in its upper part), although there is some evidence that this taxon had ecological preferences, and therefore its appearance may be facies-controlled. Moreover, in Belgium the oldest occurrence of the *Archaeopteris–Callixylon* group is *Callixylon velinense* found in strata of uppermost Givetian age. *C. velinense* has been found with *Svalbardia avelinesiana* and *S. boyi* whose status is unclear (Fairon-Demaret 1986). Taxonomic revision is clearly desirable and may well have considerable implications for zonation. A similar explanation may account for the markedly disjunct distribution of 'lepidodendroid' type stems such as *Pseudolepidodendropsis, Protolepidodendropsis* and *Leptophloeum* which are absent in Europe (except for records of the two latter form-genera near the Famennian–Frasnian boundary at Brest, France, palaeogeographically separated from Laurasia (Scotese & McKerrow 1990)) and Siberia (Meyen 1987). *Leptophloeum* itself occurs only rarely in Laurasia, being described initially from Maine, later from the Gaspé (Dawson 1871), and more recently recorded from Ellesmere Island (Scheckler *et al.* 1990). Also noteworthy are the final appearances of the Lower Devonian taxa *Sawdonia* and *D. spinaeformis*.

Scheckler (1986b) in his discussion on Banks' Biozone VI suggested its subdivision based on assemblages from New York State where aneurophytes and herbaceous lycophytes (e.g. *Colpodexylon, Haskinsia, Drepanophycus*) become scarce around the Soyea/West Falls group boundary. The upper part of the zone is dominated by *Archaeopteris* and arborescent lycophytes, the nominal form-genus itself being rare lower in the zone. *Archaeopteris* extends into the Famennian, when it is initially co-dominant with *Rhacophyton*.

Biological innovation in reproductive strategy seen within the zone is the evolution of well-defined heterospory culminating in the appearance of the 'seed' megaspore *Cystosporites*.

3.4.7 Assemblage Biozone VII — *Rhacophyton* Biozone

Chronostratigraphic range — middle Famennian to base Carboniferous.

The appearance of *Rhacophyton* in the Hampshire Formation, USA, and in the Evieux Formation, Belgium, marks the base of the zone in Banks' account, although the form-genus is recorded earlier as a petrifaction in the Givetian of New York State (Matten 1974, section 3.2.4) and also as adpressions in the Soviet Union. However, the fossils from the Russian platform were named *R. incertum*, a taxon which Andrews & Phillips (1968) thought probably unrelated to *Rhacophyton*. In Belgium itself the first appearance of *Rhacophyton* may be taphonomically constrained as there are no continental deposits older than the Evieux Formation in the Famennian.

Excellent stratigraphical control for this biozone has been provided by palynological activity relating to the Devonian/Carboniferous boundary, although Banks himself did not delimit the top of the zone now based at the 'Strunian'–Tournaisian boundary. A summary of the distribution of assemblages was given by Fairon-Demaret (1986), who demonstrated the potential of *Cyclostigma kiltorkense* as a zonal index fossil for an upper sub-zone of VII, in that apart from an Irish record it appears confined to the Strunian. Indeed, Jarvis (1990) thought it possible that the Kiltorkan record is also Devonian. *Rhacophyton* itself extends into the Lower Carboniferous, and occurs in such great abundance in pure stands at certain presumably swampy environments that it forms coal, although it also is recorded from drier habitats.

Scheckler (1986b) also considered the possibility of subdividing Biozone VII near the beginning of Tn 1a (i.e. near the base of the Strunian (Fa 2d)) with the appearance of *Cyclostigma*, but urged caution as *Rhacophyton*, *Archaeopteris* and *Cyclostigma* are not always recorded in coeval assemblages and their appearance may have been taphonomically or ecologically controlled.

Concerning bioevents, the zone is noted for the appearance of ovules and hence gymnosperms (Fa 2c) and the lowest unequivocal records of sphenophytes.

3.5 COMPARISON WITH ZONATION BASED ON PALYNOMORPHS

Since the spores of land plants can be recovered in abundance and from a much larger number of locations in both marine and continental strata, they offer far greater potential for correlation than do plant fossils. A recent comprehensive synthesis of data from the Old Red Sandstone Continent (Laurasia) and adjacent basins by Richardson & McGregor (1986) produced seventeen spore assemblage biozones in the interval (middle Wenlock to Famennian) encompassed by Banks' seven zones (Fig. 3.4), a demonstration of the finer resolution obtainable from the palynomorphs. For the same interval, there are further excellent, but more localized, spore-based biostratigraphies (e.g. Streel *et al.* 1987) and, at the top, exhaustive studies relating to the positioning of the Devonian–Carboniferous boundary (e.g. Becker *et al.* 1974). Richardson & McGregor integrated the timing of major innovations in spore morphology with their zonation, some of which coincided with their zonal boundaries.

3.5.1 *Cooksonia* Biozone

The oldest record of *Cooksonia* is roughly coincident with the base of the cf. *protophanus–verrucatus* spore Biozone, which is marked by the incoming of spores with verrucate sculpture. Crassitate trilete spores with similar sculpture have been isolated from Pridoli *Cooksonia* sporangia, as have spores belonging to *Ambitisporites*. It might therefore be anticipated that *Cooksonia* will eventually be found in Llandovery sediments, with a resulting further extension of Banks' oldest zone.

3.5.2 *Zosterophyllum* Biozone II

Zosterophyllum is first recorded near the base of the *micrornatus–newportensis* spore Biozone, and the diversification of zosterophylls within the succeeding *breconensis–zavallatus* spore Biozone. Work in progress in Cardiff on *in situ*

Series	Stages	Palynology Biozones	Plant Biozones
Upper Devonian	Tournaisian (part)	*nitidus - verrucosus*	*Rhacophyton*
		pusillites - lepidophyta	*Archaeopteris*
	Famennian	*flexuosa - cornuta*	
		torquata - gracilis	
	Frasnian	*ovalis - bulliferus*	
		optivus - triangulatus	
Middle Devonian	Givetian	*lemurata - magnificus*	*Svalbardia*
	Eifelian	*devonicus - naumovii*	*Hyenia*
		velatus - langii	
Lower Devonian	Emsian	*douglastownense -eurypterota*	*Psilophyton*
		annulatus - sextanii	
	Siegenian	*polygonalis - emsiensis*	
	Gedinnian	*breconensis - zavallatus*	*Zosterophyllum*
		micronatus - newportensis	
Pridoli	Downton-ian	*tripapillatus - spicula*	*Cooksonia*
Ludlow	Ludfordian	*libycus - poecilomorphus*	
	Gorstian		
Wenlock	Homerian	cf. *protophanus - verrucatus*	
		chulus - nanus	

Fig. 3.4 — Plant fossil and palynological biostratigraphical classifications of the Silurian and Devonian.

material is attempting to relate the diversification of spores at the base of the *micrornatus–newportensis* Biozone, at least in part, to a proliferation of rhyniophytoids. As yet, however, insufficient is known about the *in situ* spores of zosterophylls (as opposed to *Zosterophyllum*) for sensible comparison with the dispersed spore record.

3.5.3 *Psilophyton* Biozone

If the base of Banks' Biozone III is placed near the base of the Siegenian — when trimerophytes first appear — then there is no coeval 'event' in the *polygonalis–emsiensis* zone to coincide with it (see also Richardson & McGregor 1986, p. 4).

3.5.4 *Hyenia* Biozone

At least one taxon in Banks' *Hyenia* zone now extends into the Emsian, and is thus closer to the base of the *douglastownense–eurypterota* spore Biozone, but again more needs to be learned of the spores of such taxa as *Pseudosporochnus*, *Hyenia* and *Calamophyton*.

3.5.5 *Svalbardia* Biozone

As McGregor & Richardson indicated, the base of Banks' zone more or less coincides with the base of the *lemurata–magnificus* spore Biozone. This is not surprising in that *Geminospora lemurata* has been isolated from *Svalbardia polymorpha* (Høeg 1942). It is perhaps slightly puzzling that the base of Banks' zone is lower than the spore zone, but the age of Høeg's material is equivocal. On the other hand, this may well provide indirect evidence that *Actinoxylon* is a preservational state of *Svalbardia* (but see Scheckler 1978).

3.5.6 *Archaeopteris* Biozone

Its base occurs in the middle of the *optivus-triangulatus* Biozone, but coincides with the incoming of monolete spores and those with a pseudo-saccate zona. The microspores of *Archaeopteris* itself are assigned, as in *Svalbardia*, to *Geminospora*. In the more local Oppel Zone scheme developed for the Ardenne–Rhineland region (Streel *et al.* 1987) the base of Banks' Zone VI occurs within the TCo Oppel Zone, which is roughly equivalent to Richardson & McGregor's spore Assemblage zone and is based on the joint occurrence of *Chelinospora concinna* and *Cirratriradites jekhowskyi* as well as *Samarisporites triangulatus*.

3.5.7 *Rhacophyton* Biozone

The base occurs in the upper part of the *torquata–gracilis* spore Biozone (Richardson & McGregor) but is more or less coincident with the traditional subdivision Fa 2b/Fa 2c of the Famennian and with the GF-VCo Oppel Zone boundary of Streel *et al.* (1987), based on the Ardenne–Rhineland regions. A summary of the distribution of plant fossil assemblages in relation to the various zonation schemes is given in Fairon-Demaret (1986).

3.6 PALAEOPHYTOGEOGRAPHICAL ANALYSIS

Two approaches have been employed for the Lower Devonian — (a) that of mere scrutinization of species lists in somewhat subjective attempts to detect concentrations of endemics or consistent patterns in distribution, and (b) that involving multivariate statistical analyses. Using the latter, North American workers have produced the most recent and objective analysis. The three sets of conclusions and interpretations in Table 3.1 demonstrate the evolution and refinement of their approach. Thus, for example, in the initial study, which involved the whole of the Lower Devonian, the results are clearly time-transgressive and as the authors themselves suggested were not a correct representation of palaeogeographical patterns.

Table 3.1 — Three recently produced palaeophytogeographical classifications for the Devonian

	Authors/stratigraphy	Palaeokingdoms	Palaeoareas	Climate
1.	Ziegler *et al.* 1981 L. Dev^{n}: undifferentiated	Gondwana		Cool south temperate
		Western N. America		Temperate–subhumid
		Equatorial	Appalachian Western European Eastern Equatorial	
2.	Raymond, Parker & Barrett 1985 Mid Siegenian–Emsian	S. Gondwana		Temperate humid to subhumid, cold seasonal climate
		Australia (23)		subtropical humid–wet perhaps mildly seasonal climate
		Equatorial & low latitude	Europe & Wales	Tropical–subtropical wet
			America	Tropical–subtropical wet
			China	Tropical–subtropical wet and humid
			Kazachstan	Tropical–subtropical wet
			Siberia	Wet and humid tropical to subtropical northern temperate subhumid
	Gedinnian–L. Siegenian	Australia Rest of world		
3.	Raymond 1987	Australia		Mild seasonal with dry or dry and cool season
		Kazachstan–north Gondwana		Tropical to mildly seasonal with a dry or dry and cool season
		Equatorial mid-latitude	Siberia–north Laurasia	Mildly seasonal with a dry or dry and cool season
		China		Wet tropical
		South Laurasia		Wet tropical

The phytochorial units have been modified to make them compatible with the rest of the book (see section 1.5.2).

For the mid-Siegenian to Emsian, Raymond *et al.* (1985) distinguished three major units, South Gondwana, Australia and an equatorial and low-latitude unit which was further subdivided along a north-west–south-east gradient into five integrading units. Differences between Europe and North America may again be a consequence of age as the localities were in relatively close proximity towards the end of the Early Devonian. The unexpected similarities between China and North Gondwana can perhaps be explained by the exceedingly optimistic form-generic names applied to the very fragmentary and often sterile fossils from the western Sahara. The China/Germany anomaly is harder to account for. Most evidence indicates that for much of the Lower Devonian, German assemblages have most in common with Belgian, while recent research from China indicates a plant fossil assemblage rich in endemics. The clustering of localities from Spitsbergen, Wyoming and Arctic Canada — the north-west fringes of Laurentia — with Siberia, again may reflect the weaknesses of the data. Tanner's largely unpublished PhD thesis on Wyoming suggests closer similarities with the rest of Laurentia. The Spitsbergen record is fragmentary, and the Bathurst Island assemblage may be taphonomically impoverished (see section 3.2.3).

Very similar results were obtained with an alternative approach by Raymond (1987), who attempted to eliminate regional taxonomic bias by basing her analysis on the distribution of certain morphological characters. As a consequence only three subunits were distinguished in an equatorial mid-latitude unit. North America and Europe are united, China stands alone and there is again a Siberian–north Laurasia connection. Kazakhstan is distinguished as a major, separate unit, united with north Gondwana.

More traditional analyses by Petrosian (1968) and Edwards (1973) were undertaken soon after the beginnings of the resurgence of interest in early land plants and encompassed the whole of the Devonian. In the early part, Edwards found no convincing evidence for distinctive phytochoria except between Gondwana and localities now occurring in the northern hemisphere and Australia. More critical analysis of the ages of Gondwana occurrences (e.g. Edwards 1990b) questions their existence in South Africa (but see Rayner 1988) and South America, except for very recent collections from Argentina (work in progress), while Australia clearly had a distinctive flora from Ludlow times (Silurian) onwards. Although *Yarravia* and *Hedeia* are recorded elsewhere, their identification is not convincing, and *Baragwanathia* is not reported from Laurasia below the upper Emsian. Most noticeably absent from Australia is the very widespread *Drepanophycus* (even *sensu lato*). Li & Edwards (in press) commented on the high percentage of endemics in Siegenian–Emsian assemblages from South Yunnan Province (China).

More recently, Edwards (1990a) noted the distinctive nature of Kazakhstan and Siberian assemblages in upper Silurian and presumed Gedinnian strata respectively. Stepanov (1975) had distinguished two 'phytocomplexes' in the Emsian of the Kuznetsk Basin, western Siberia. Mere scrutiny of his species does suggest some similarities with Laurasia, although there are endemics too. However, more detailed analysis indicates that quite complex fertile branching systems should not be assigned to either *Cooksonia hemisphaerica* or *C. pertoni* (*?Renalia*) and that the identification of the spiny axes as various species of *Psilophyton* is also conjectural in the *Margophyton*/*Psilophyton burnotense*/*Sawdonia* controversy.

Thus more recent work reinforces, to a certain extent, the three phytochoria recognized in the Lower Devonian of the Soviet Union by Petrosian (1968), although Meyen (1987) found it impossible to distinguish phytochoria either in the USSR or globally because of inadequacies of the record and the nature of land vegetation.

	Greenland, Spitzbergen, Bear Island	Russian Platform	Belgium, Germany, Northern France	Southern British Isles	Scotland, Norway
Protolepidodendropsis	+				
Sublepidodendron	+				
Cephalopteris	+				
Pseudolepidodendropsis	+				
Protocephalopteris	+	+			
Enigmophyton	+	+			
Pseudobornia	+	+			
Sphenophyllum	+	+	+		
Salopella	+	?		+	
Cyclostigma	+		+	+	
Archaeopteris	+	+	+	+	
Svalbardia	+	+	+		+
Hyenia	+	+	+		+
Psilophyton/Dawsonites	+	+	+	+	+
Zosterophyllum	+	?	+	+	+
Pertica		+			
Stockmansella		+	+		
Barinophyton		+	+		
Calamophyton		+	+		
Lycopodites		+	+		
Aneurophyton		+	+		
Callixylon		+	+		
Rhacophyton		+	+		
Gosslingia		+	+	+	

Fig. 3.5 — Geographical distribution of Silurian and Devonian form-genera in European Laurasia. Russian Platform localities include Podolia, Donbass, the Baltic States, Kola Peninsula, Timan and the Volga–Urals region. Southern British Isles includes South Wales and the Borders, Devon and southern Ireland.

In summary for the Lower Devonian, there appears to be agreement in the documentation of Australian, Chinese, Siberian, Kazakhstan and Laurasian phytochoria, although subdivision of Laurasia is more controversial. As far as the Middle and Upper Devonian are concerned little has been published in terms of global palaeophytogeography, although some authors (Petrosyan 1968, Scheckler 1986c) have emphasized the distinct nature of the Siberian and Kazakhstan assemblages. There have been no statistical analyses, such as those undertaken in the Lower Devonian. The generalization that there was then a worldwide uniform flora with no readily identifiable phytochoria may reflect an incomplete and inadequately analysed record.

On the other hand, it may relate to a global biological trend, in that marine invertebrate faunas appear to show a change from marked provincialism in the Early Devonian to high cosmopolitanism in the Late Devonian, and a possible reduction in the climatic gradient during those times has been postulated (Boucot 1988). Scheckler (1986c) suggested that a decrease in diversity of Frasnian plant fossil

	Greenland, Spitzbergen, Bear Island	Russian Platform	Belgium, Germany, Northern France	Southern British Isles	Scotland, Norway
Taeniocrada		?	+	?	
Barrandeina		+			+
Cooksonia		+		+	+
Rellimia		+	+		+
Pseudosporochnus		+	+		+
Drepanophycus		+	+	+	+
Sawdonia		?	+	+	+
Rebuchia			+		
Cladoxylon			+		
Protolepidodendron			+		
Leclercqia			+		
Triloboxylon			+		
Tetraxylopteris			+		
Honsleria			+		
Serrulacaulis			+		
Eviostachya			+		
Renalia			+	+	
Sphenoptreris			+	+	
Sporogonites			+	+	+
Uskiella				+	
Lepidodendropsis				+	
Hydrasperma				+	
Pectinophyton					+

Fig. 3.5 — Geographical distribution of Silurian and Devonian form-genera in European Laurasia. Russian Platform localities include Podolia, Donbass, the Baltic States, Kola Peninsula, Timan and the Volga–Urals region. Southern British Isles includes South Wales and the Borders, Devon and southern Ireland.

assemblages in North America resulted from low rates of origination, rather than a Frasnian extinction event as seen in the marine record. This may have reflected a gradual change of climate rather than an abrupt event.

Meyen (1987) commented on the distribution of records of the larger (arborescent) lycopsids in both the Middle and Upper Devonian. In the Middle Devonian, occurrences of such plants are noted in Kazakhstan, Spitsbergen, Libya, North America and China, although he says that 'it is still unclear whether the localities can be united into an independent phytochorion, or whether the occurrence here of lepidophytes with rather thick stems was controlled by local facies conditions'. He also noted the absence of the common European form-genera *Hyenia* and *Calamophyton* from China. Reviews of Chinese assemblages suggest considerable diversity (e.g. Dou & Sun 1983). Many contain a high percentage of leafless or decorticated lycophyte stems, and these, together with specimens assigned to Laurasian taxa, require reassessment before their palaeophytogeographical significance can be evaluated (see e.g. Li 1990). In the Upper Devonian, Meyen perceived little distinct

	Maine, Ontario, Gaspé, New Brunswick	New York State	Ohio, Virginia, West Virginia, Indiana, Kentucky	Wyoming	West Alaska Coast	Alaska, Canadian Arctic, Alberta
Baragwanathia	+					
Zosterophyllum	+					
Renalia	+					
Barrandeina	?					
Cyclostigma	?					
Calamophyton	+	+				
Tetraxylopteris	?	+				
Pertica	+		+			
Rhacophyton	?	?	+			
Barinophyton	+	+	+			
Leptophloeum	+					+
Archaeopteris	+	+	+			+
Sawdonia	+	+		+		
Leclercqia	+	+		+		
Psilophyton/Dawsonites	+			?		
Drepanophycus	+	+			+	+
Cooksonia		+				
Hyenia		+				
Rellimia		+				
Pseudosporochnus		+				
Triloboxylon		+				

	Maine, Ontario, Gaspé, New Brunswick	New York State	Ohio, Virginia, West Virginia, Indiana, Kentucky	Wyoming	West Alaska Coast	Alaska, Canadian Arctic, Alberta
Aneurophyton		+				
Haskinsia		+				
Colpodexylon		+				
Xenocladia		+				
Archaeosigillaria		+				
Serrulacaulis		+				
Svalbardia		+				?
Astralocaulis		+				
Lepidosigillaria		+				
Lepidostrobus		+				
Arachnoxylon		+	+			
Cladoxylon		+	+			
Iridopteris		+	+			
Callixylon		+	+			
Protobarinophyton			+			
Sphenopteris			+			
Sphenophyllum			+			
Gosslingia				?		
Rebuchia				+		+
Protolepidodendropsis						?
Pseudobornia						+

Fig. 3.6 — Geographical distribution of Silurian and Devonian form-genera in American Laurasia.

	Australia	Libya, Morocco, Algeria	South Africa	Brazil	Venezuela	Antarctica
Salopella	+					
Baragwanathia	+					
Zosterophyllum	+					
Sporogonites	+					
Barinophyton	+					
Psilophyton/Dawsonites	?					
Leclercqia	+					
Archaeopteris	+					
Lepidosigillaria	?	+				
Leptophloeum	+		+			
Uskiella	+			?		
Astralocaulis	+				+	
Cooksonia		+				
Lepidodendropsis		+				
Pseudolepidodendropsis		+				
Archaeosigillaria		+	+	+		
Haskinsia		?			+	?
Pseudosporochnus					+	
Tetraxylopteris					+	
Colpodexylon					+	
Serrulacaulis					+	
Protolepidodendropsis						+

Fig. 3.7 — Geographical distribution of Silurian and Devonian form-genera in Gondwana.

regional variation in plant fossil assemblages especially in those of Banks' Zone VI, although his observations on the distribution of *Leptophloeum* and other lycopsids have been noted elsewhere (section 3.4.6). In Banks' Zone VII, however, Meyen saw the beginning of the differentiation that led to the separation of the distinct Angara Palaeokingdom (Meyen 1982) of Siberia in the Lower Carboniferous (see section 4.3).

The small amount of information on Middle and Upper Devonian Gondwana assemblages does little to conflict with the notion of a 'worldwide' flora during these times, for instance the occurrences of *Leclercqia* and *Astralocaulis* in Australia. The discovery of a diverse assemblage of plant fossils in upper Givetian/lower Frasnian strata in north-western Gondwana (i.e. Venezuela — Edwards & Benedetto 1985) supports this. The assemblage, currently being studied in Cardiff, includes such plants as herbaceous lycophytes (a few species referable to *Colpodexylon*, *Haskinsia* and *?Archaeosigillaria*), *Pseudosporochnus*, *Tetraxylopteris*, and *Serrulacaulis*, and seems to bear a close affinity to assemblages from New York State. However, whether this assemblage can be taken to be representative of Gondwana as a whole is questionable, given the close proximity of New York State and Western Venezuela in some palaeogeographical reconstructions (e.g. Heckel & Witzke 1979), or the projection of north-western South America further north than adjoining Gondwana areas (Kent & van der Voo (1990)). The distribution of plant fossils in Gondwana prompted Edwards (1990b) to 'very tentatively suggest that the high latitude (Gondwana) vegetation was dominated by herbaceous lycophytes in the Middle to

	Brest, Poland, Czechoslovakia	Romania, Bulgaria	Kazakhstan, North West Xingiang	Western Siberia	North China, Japan, Vladivostok	South China, Vietnam
Cooksonia	+					
Sporogonites	+					
Taeniocrada	+			?		
Pseudosporochnus	+	+	+	+		+
Aneurophyton	?	+		+		
Archaeopteris	+		+	+	+	+
Leptophloeum	+		+		+	+
Drepanophycus	+		+	+		+
Psilophyton/Dawsonites	+		?	+		
Barrandeina	+		+		+	+
Protolepidodendropsis	+				+	
Sublepidodendron	+				+	+
Calamophyton		+	+			
Sphenophyllum		+	+	+	+	
Salopella			?			
Lycopodites			+			
Xenocladia			+			
Enigmophyton			+			
Eviostachya			+			
Rhacophyton			+	+		

	Brest, Poland, Czechoslovakia	Romania, Bulgaria	Kazakhstan, North West Xingiang	Western Siberia	North China, Japan, Vladivostok	South China
Pseudobornia			+	+	+	
Svalbardia			+	+	+	
Cyclostigma			+	+	+	+
Zosterophyllum			+	+		+
Archaeosigillaria			+		+	+
Lepidodendropsis			+		+	+
Renalia				?		
Rebuchia				+		
Sawdonia				?		
Protobarinophyton				+		
Rellimia				+		
Pectinophyton				+		
Pseudolepidodendropsis				+		
Protocephalopteris				+	+	
Hyenia				+	+	
Calixylon					+	
Colpodexylon					+	+
Sphenopteris					+	+
Lepidosigillaria					+	+
Lepidostrobus					+	?

Fig. 3.8 — Geographical distribution of Silurian and Devonian form-genera in other areas.

early Late Devonian and was of lower diversity than the vegetation of lower latitudes'.

Palynological records from Middle and Upper Devonian sediments outnumber those of plant fossils, but their application to palaeophytogeographical analysis is still in its infancy. Streel *et al.* (1990) remarked that the evolution of heterospory and the seed habit would have imposed restrictions on dispersal and hence promoted endemism. Their studies on Givetian and Frasnian sediments from Laurasia and western Gondwana suggested a uniform vegetation in southern Laurasia and western Gondwana during this period with the possible distribution of an equatorial belt in northern Laurasia in the Frasnian, but Fairon-Demaret (loc. cit.), could see no clear-cut differences in coeval plant fossil assemblages. However, an abstract published by Scheckler *et al.* (1990) with a summary of Eifelian to middle-upper Frasnian assemblages from Ellesmere, Arctic Canada (palaeolatitude 30°N), indicates that these are less diverse than in tropical palaeolatitudes but, unlike the higher latitudes of Gondwana, lack herbaceous lycophytes.

3.7 CONCLUDING REMARKS

In that outside the Silurian and Lower Devonian the authors have limited working knowledge of the assemblages under scrutiny, they have been reluctant to present any new conclusions. It has been the aim to present data relevant to the application of plant fossils to stratigraphy and to summarize progress on palaeophytogeographical analyses. In the eleven years since Banks formulated his zonation scheme, the overall pattern has changed very little, but there have been disappointingly few attempts to attain greater precision (i.e. subdivision). This may result from the fact that there are many form-genera which, although geographically widespread, are also of long stratigraphical range (e.g. Eifelian to lower Frasnian) and where more precise dating may depend on the preservation of certain key form-genera, these may show ecological preferences and hence be facies-controlled (e.g. *Rhacophyton*). As an example of the former the authors can cite their own experience in Venezuela, where an extensive plant fossil assemblage occurs in predominantly fluvial sediments. The lowest plant-bearing horizon contains the lycopods *Haskinsia* and *Colpodexylon*, the zosterophyll *Serrulacaulis*, and the cladoxylalean *Pseudosporochnus*, as well as possible zosterophylls and progymnosperms not yet identified. Beds at least 150 m higher in the sequence contain an assemblage including a separate species of *Colpodexylon*, as well as *Tetraxylopteris* and *Astralocaulis*.

With the exception of the rare *Astralocaulis*, all these named plants may have coexisted from all but the earliest Givetian through till early Frasnian times. Indeed, all these form-genera are found in the Frasnian Oneonta Formation in New York State (Banks 1966). However, the Venezuelan assemblage contains a notable lack of *Archaeopteris*. It has been pointed out already that the plant fossil zones of Banks are marked by the recognition of critical form-genera amongst other long-lived varieties. In the case of the Venezuela assemblages the lack of *Archaeopteris* may reflect ecological control rather than a true age constraint.

The Venezuelan plants are in fact constrained by the middle–late Givetian age of a brachiopod assemblage some 50 m below the lowest plant-bearing horizon, and a late Early Carboniferous–early Late Carboniferous assemblage of brachiopods and bivalves some 300 m above. The determination from plant fossils is thus supported by the independent evidence. Perhaps the failure to date the sediments by the plant

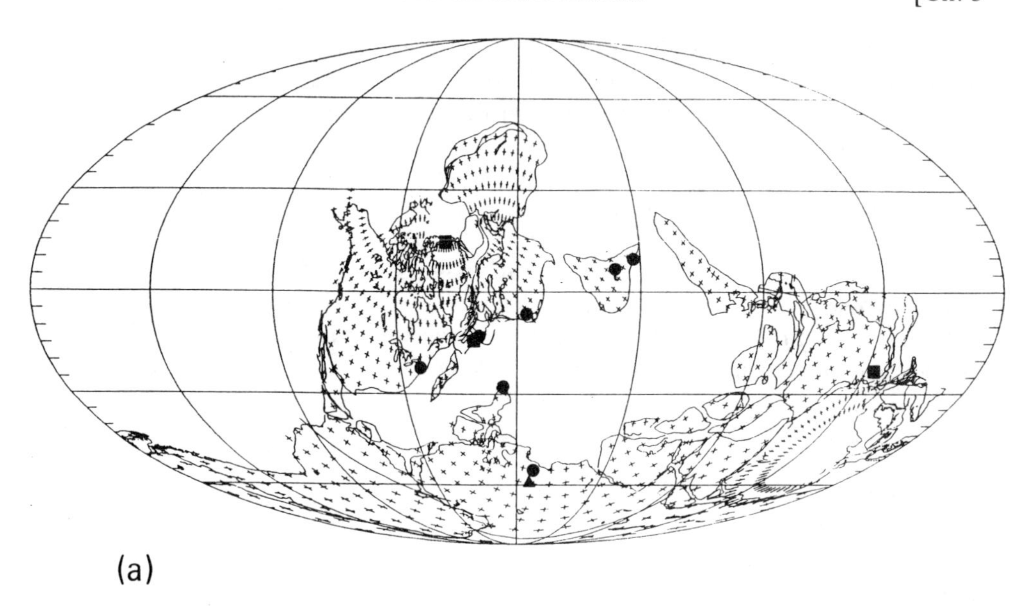

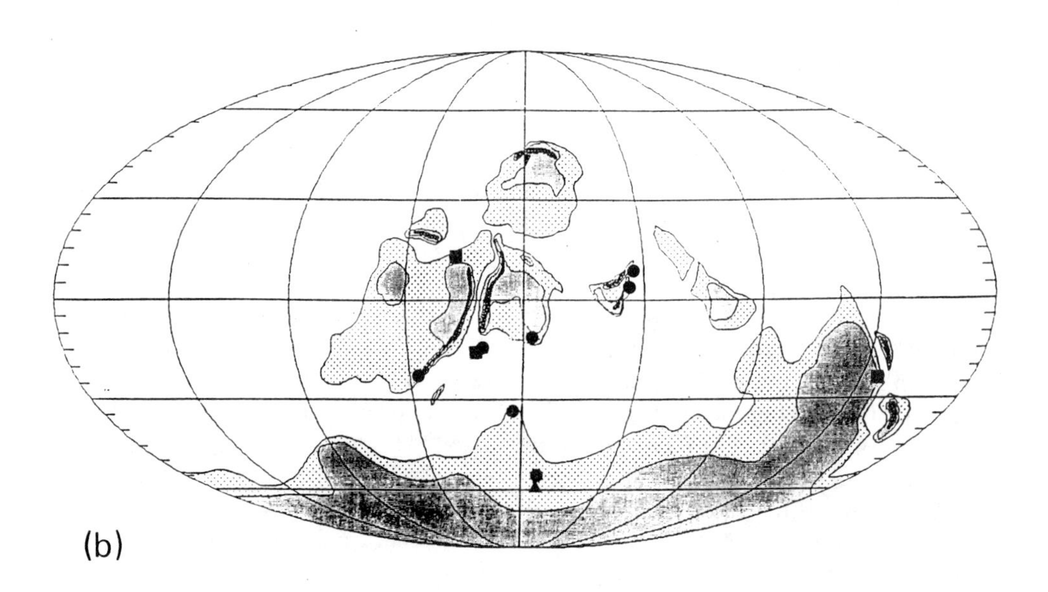

Fig. 3.9 — Main Silurian plant fossil localities, plotted on Ludlow palaeogeographical (a) and Wenlock land/sea (b) maps. Base maps from Scotese & McKerrow (1990). Squares, Ludlow sites; circles, Pridoli sites; triangles, Silurian sites of uncertain series position. Shading: grey stipple, land; light stipple, shallow sea; dense shading, mountains and island arcs.

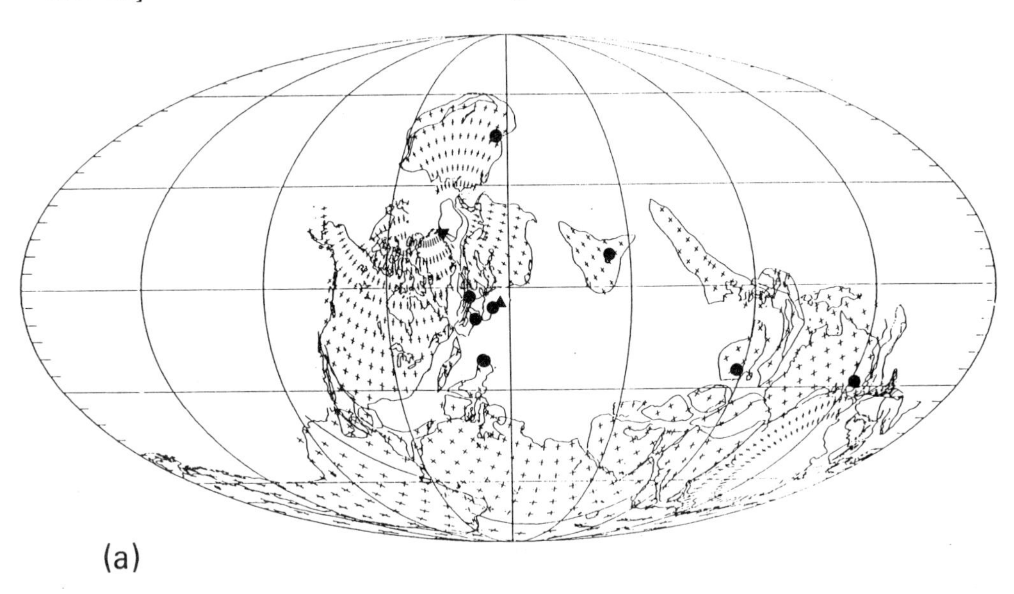

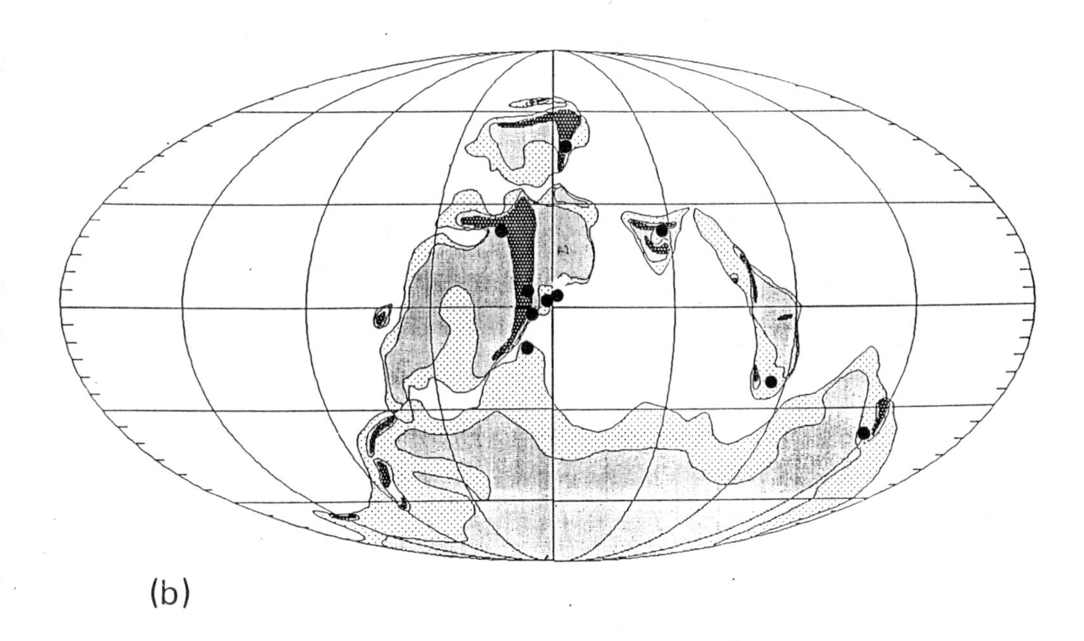

Fig. 3.10 — Main Gedinnian plant fossil localities, plotted on Gedinnian palaeogeographical (a) and Emsian land/sea (b) maps. Base maps from Scotese & McKerrow (1990). Shading as in Fig. 3.9.

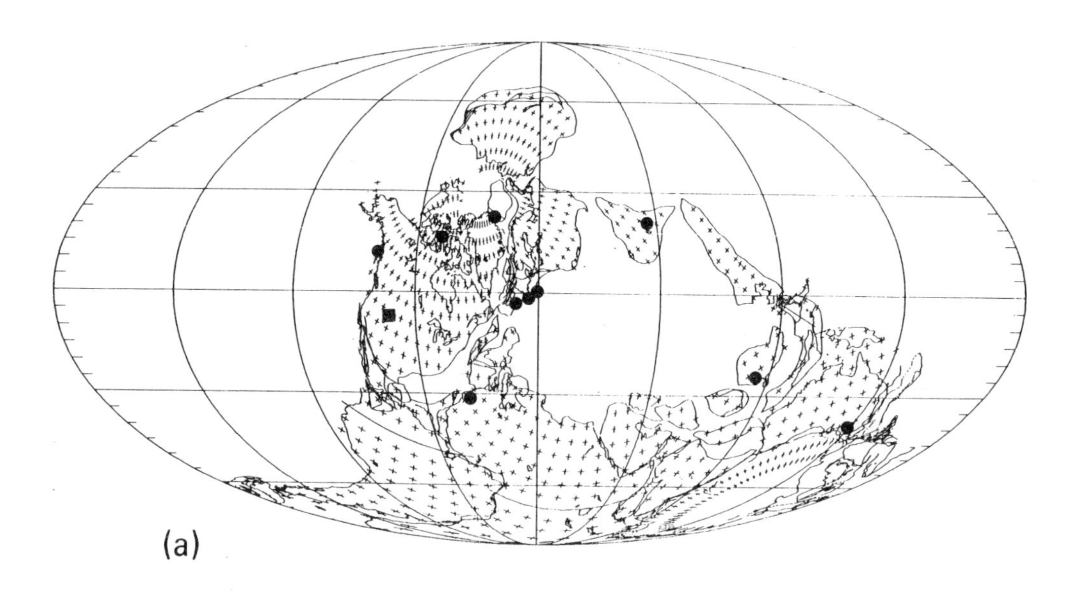

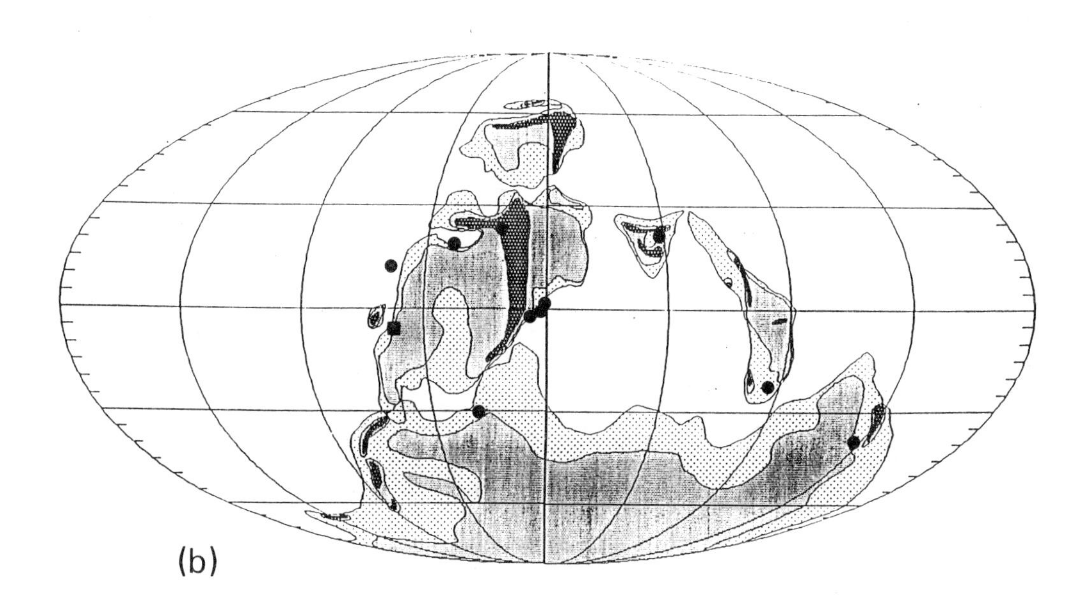

Fig. 3.11 — Main Siegenian plant fossil localities, plotted on Gedinnian palaeogeographical (a) and Emsian land/sea (b) maps. Base maps from Scotese & McKerrow (1990). Circles, well-dated sites; squares, Lower Devonian probably Sigenian sites. Shading as for Fig. 3.9.

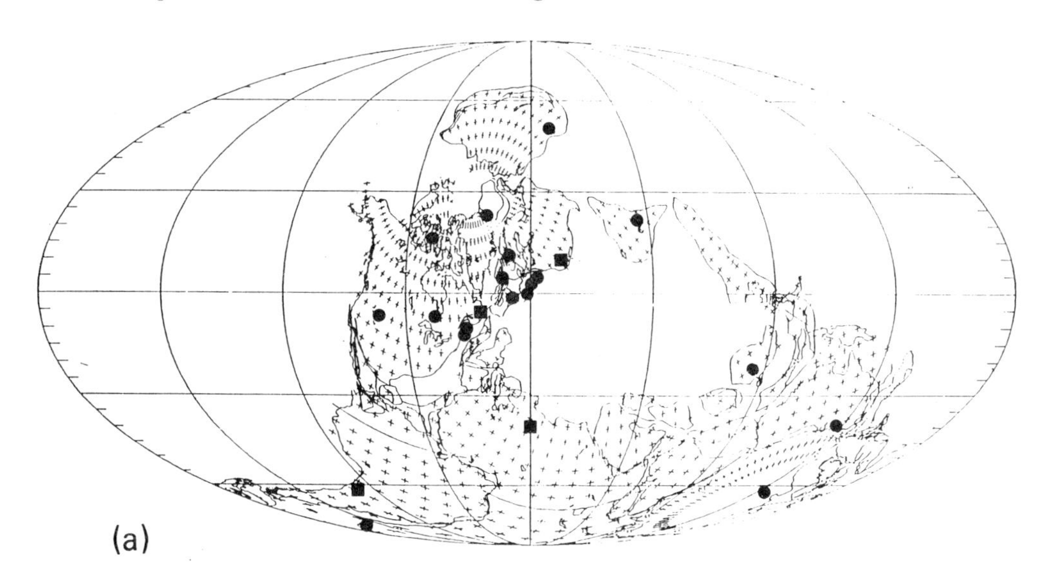

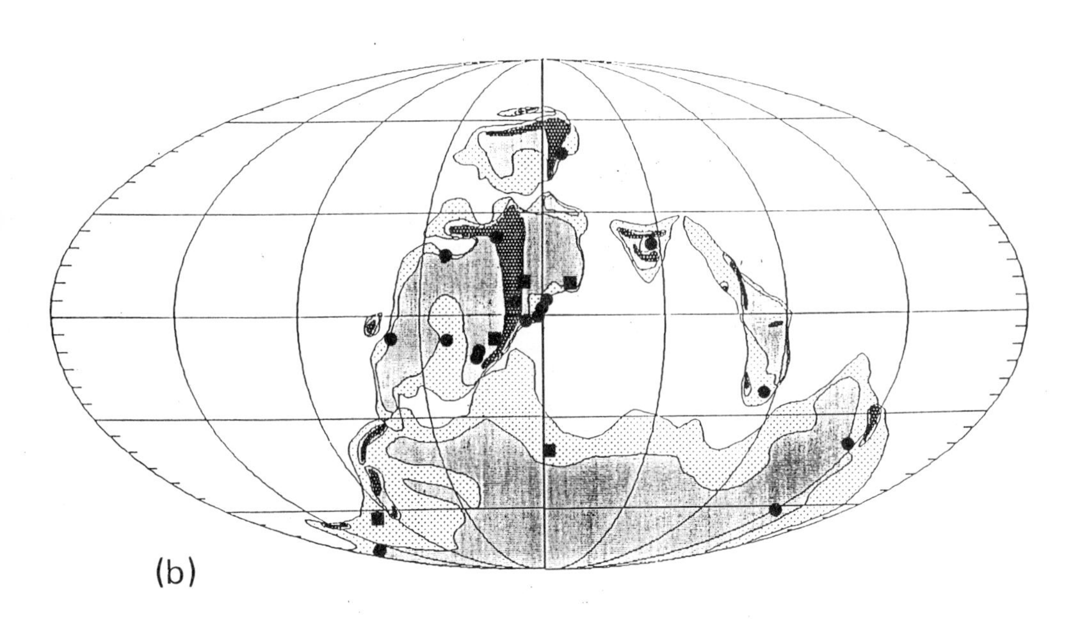

Fig. 3.12 — Main Emsian plant fossil localities, plotted on Gedinnian palaeogeographical (a) and Emsian land/sea (b) maps. Base maps from Scotese & McKerrow (1990). Circles, well-dated sites; squares, indetermine Lower Devonian. Shading as for Fig. 3.9.

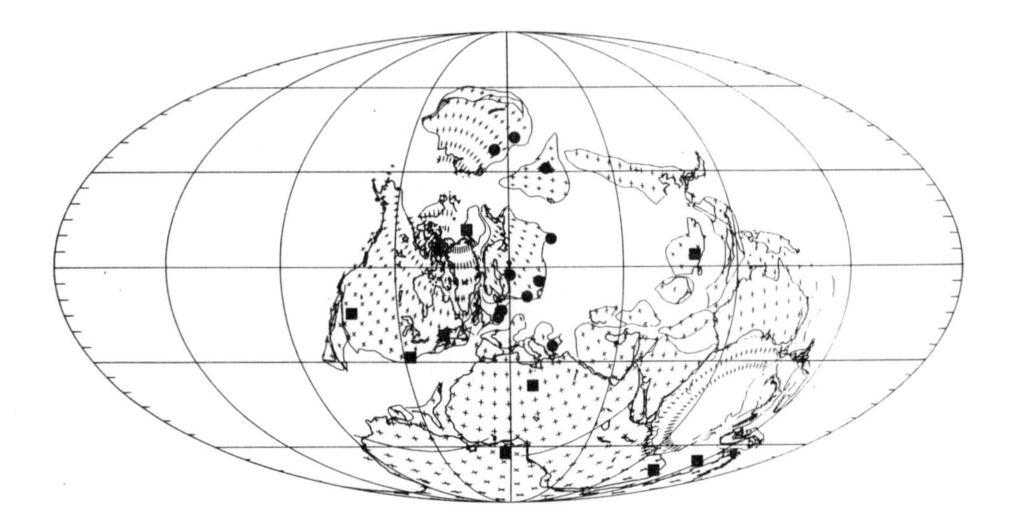

Fig. 3.13 — Main Eifelian plant fossil localities, plotted on Givetian palaeogeographical map. Base map from Scotese & McKerrow (1990). Circles, well-dated sites; squares, Middle Devonian probably Eifelian sites.

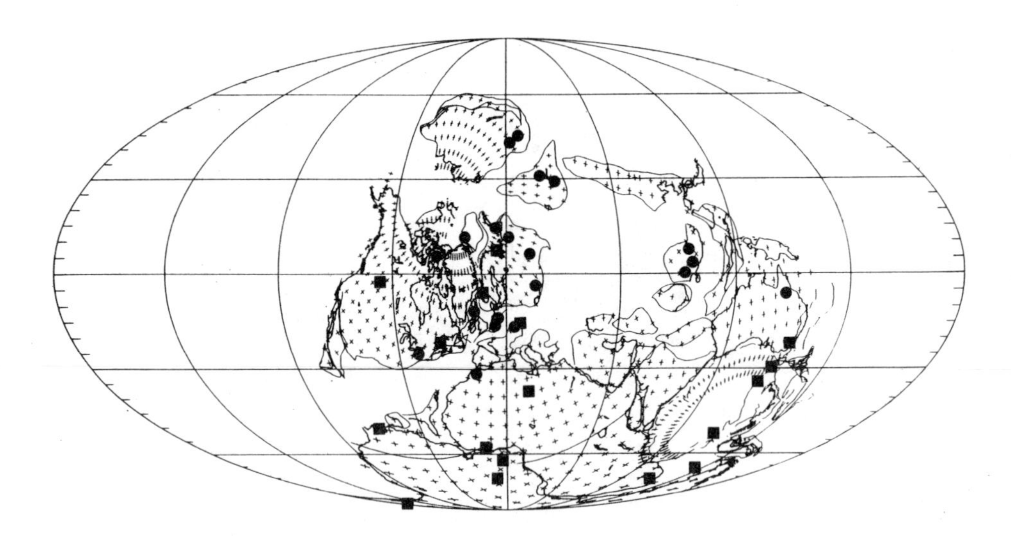

Fig. 3.14 — Main Givetian plant fossil localities, plotted on Givetian palaeogeographical map. Base map from Scotese & McKerrow (1990). Circles, well-dated sites; squares, Middle Devonian probably Givetian sites.

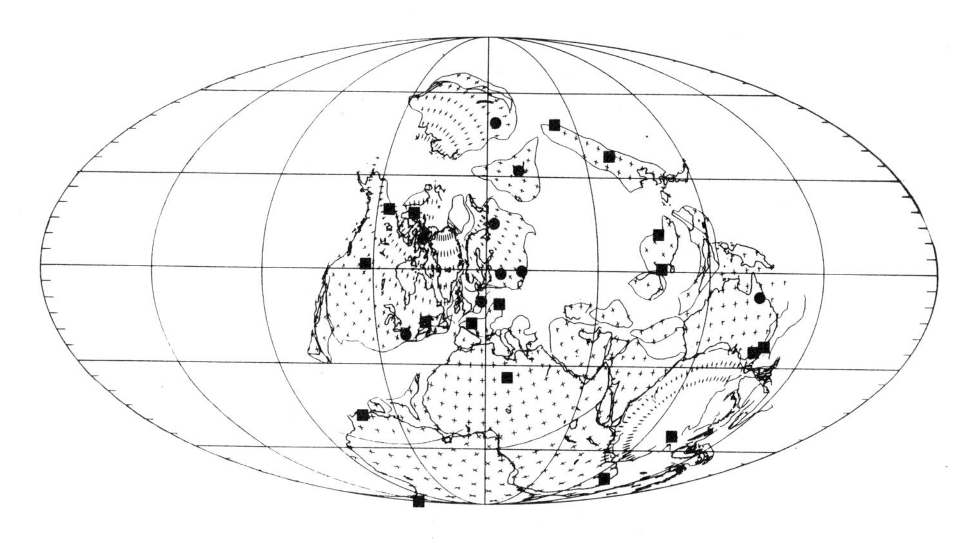

Fig. 3.15 — Main Frasnian plant fossil localities, plotted on Famennian palaeogeographical map. Base map. from Scotese & McKerrow (1990). Circles, well-dated sites; squares, Upper Devonian probably Frasnian sites.

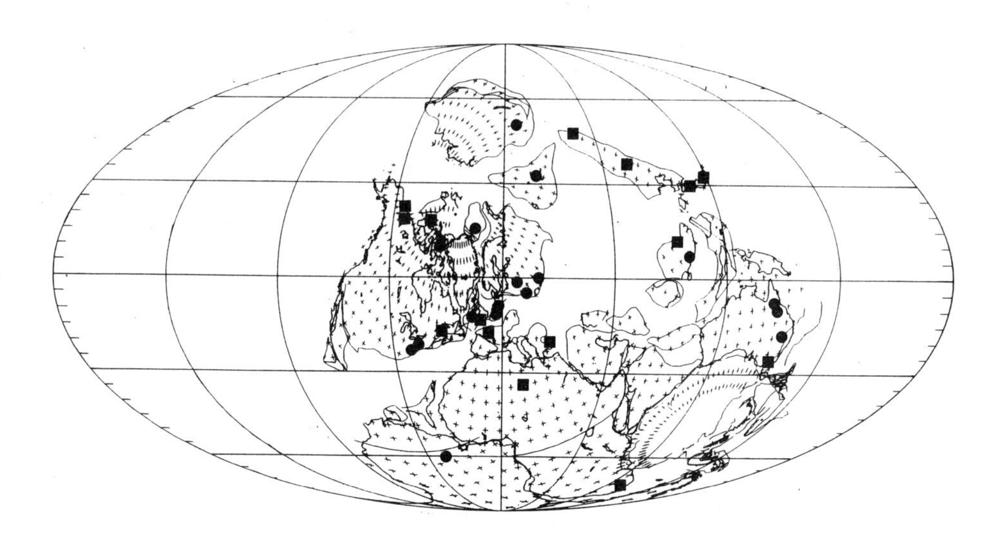

Fig. 3.16 — Main Famennian plant fossil localities, plotted on Famennian palaeogeographical map. Base map from Scotese & McKerrow (1990). Circle, well-dated sites; squares, Upper Devonian probably Famennian sites.

fossils to the level of Banks' zone, even to a stage, is a realistic assessment of the state of the art of Devonian plant biostratigraphy.

REFERENCES

Andrews, H. N. & Phillips, T. L. (1968) *Rhacophyton* from the Upper Devonian of West Virginia. *J. Linn. Soc. (Botany)* **61** 384, 37–64.

Arber, E. A. N. (1921) *Devonian floras.* University Press, Cambridge.

Ball, H. W. & Dineley, D. L. (1961) The Old Red Sandstone of Brown Clee Hill and the adjacent area. 1. Stratigraphy. *Bull. Br. Mus. Nat. Hist. (Geol.)* **5** 175–242.

Banks, H. P. (1966) Devonian flora of New York State. *Empire State Geogram* **4** 10–24.

Banks, H. P. (1980) Floral assemblages in the Siluro-Devonian. In: Dilcher, D. L. & Taylor, T. N. (eds) *Biostratigraphy of fossil plants.* Dowden, Hutchinson & Ross, Stroudsburg, pp. 1–24.

Banks, H. P. (1981) Time of appearance of some plant biocharacters during Siluro-Devonian time. *Can. J. Bot.* **59** 1292–1296.

Banks, H. P., Bonamo, P. B. & Grierson, J. D. (1972) *Leclercqia complexa* gen. et sp. nov., a new lycopod from the late Middle Devonian of eastern New York. *Rev. Palaeobot. Palynol.* **14** 19–40.

Becker, G., Bless, M. J. M., Streel, M. & Thorez, J. (1974) Palynology and ostracode distribution in the Upper Devonian and basal Dinantian of Belgium and their dependence on sedimentary facies. *Med. Rijks. Geol. Dienst N.S.* **25** 9–99.

Bonamo, P. M., Banks, H. P. & Grierson, J. D. (1988) *Leclercqia Haskinsia*, and the role of leaves in the delineation of Devonian lyopid genera. *Bot. Gaz.* **149** 222–239.

Boucot, A. J. (1988) Devonian biogeography; an update. In: McMillan, N. J., Embry, A. F. & Glass, D. J. (eds). *Devonian of the World.* Vol. III. Memoir 14, Canadian Society of Petroleum Geologists, Calgary, pp. 211–227.

Cai C.-Y. (1981) On the occurrence of *Archaeopteris* in China. *Acta Palaeont. Sinica,* **20** 75–79.

Cai, C.-Y. (1989) Two *Callixylon* species from Upper Devonian of Junggar Basin, Xinjiang. *Acta Palaeont. Sinica.* **28** 571–578.

Cai C.-Y. & Qin H.-Z. (1986) First discovery of a stem with internal structure referable to *Leptophloeum rhombicum* from the Upper Devonian, Xinjiang. *Acta Palaeont. Sinica* **25** 516–524.

Cai C.-Y, Wen Y.-G. & Chen P. Q. (1987) *Archaeopteris* florula from Upper Devonian of Xinhui county, central Guangdong and its stratigraphical significance. *Acta Palaeont. Sinica* **26** 55–64.

Chaloner, W. G. & Sheerin, A. (1979) Devonian macrofloras. *Special Papers in Palaeontology,* **23**, 145–161.

Churkin, M., Eberlein, G. D., Hueber, F. M. & Mamay, S. H. (1969) Lower Devonian land plants from graptolitic shale in south-eastern Alaska. *Palaeontology* **12** 559–573.

Cuerda, A., Cingolani, C., Arrondo, O., Morel, E. and Ganuza, D. (1987) Primer registro de plantas vasculares en la Formacion Villavicencio, Precordillera de Mendoza, Argentina. *IV Congr. Lantinoamer. Paleont* (Bolivia, 1987) **1** 179–183.

Daber, R. (1971) *Cooksonia*: one of the most ancient psilophytes — widely distributed, but rare. *Botanique* **2** 35–39.

Dawson, J. W. (1871) Report on the fossil land plants of the Devonian and Upper Silurian formations of Canada. *Geol. Surv. Canada Publ.* **428** 1–92.

Dou Y.-W. & Sun Z.-H. (1983) Devonian plants. In: *Palaeontological Atlas of Xinjiang, Vol. II. Late Palaeozoic Section.* Geological Publishing House, Beijing.

Dou Y.-W. & Sun Z. H. (1985) On the late Palaeozoic plants in northern Xinjiang. *Acta Geol. Sinica* **59** 1–10 [In Chinese].

Douglas, J. G. & Lejal-Nicol, A. (1981) Sur les premiéres flores vasculaires terrestres datée du Silurian: une comparaison entre la 'Flore a *Baragwanathia*' d'Australie et la 'Flore a Psilophytes et Lycophytes' du Nord. *C. R. Acad. Sci., Paris* **292** 685–687.

Edwards, D. (1973) Devonian floras. In: Hallam, A. (ed.). *Atlas of palaeobiogeography.* Elsevier, Amsterdam, pp. 105–115.

Edwards, D. (1990a) Constraints on Silurian and Early Devonian phytogeographic analysis — based on megafossils. In: McKerrow, W. S. & Scotese, C. R. (eds). *Palaeozoic palaeogeography and biogeography.* The Geological Society, London (Memoir No. 12), pp. 233–242.

Edwards, D. (1990b) Silurian–Devonian paleobotany: problems, progress and potential. In: Taylor, T. N. & Taylor, E. L. (eds). *Antarctic paleobiology. Its role in the reconstruction of Gondwana.* Springer Verlag, New York, pp. 89–101.

Edwards, D. (in press) Aspects of research on *in situ* spores in early land plants. *Proc. geol. Ass.*

Edwards, D. & Benedetto, J. L. (1985) Two new species of herbaceous lycopods from the Devonian of Venezuela with comments on their taphonomy. *Palaeontology* **28** 599–618.

Edwards, D. & Davies, M. S. (1990) Interpretation of early land plant radiations: facile adaptationist guesswork or reasoned speculation? In: Taylor, P. D. & Larwood, G. P. (eds). *Major evolutionary radiations. Systematics Association Special Volume No. 42,* Clarendon Press, Oxford, pp. 351–376.

Edwards, D. & Fanning, U. (1985) Evolution and environment in the late Silurian–early Devonian: the rise of the pteridophytes. *Phil. Trans. R. Soc. Lond. B* **309** 147–165.

Edwards, D., Feehan, J. & Smith, D. G. (1983) A late Wenlock flora from Co. Tipperary, Ireland. *Bot. J. Linn. Soc.* **86** 19–36.

Edwards, D., Kenrick, P. & Carluccio, L. M. (1989) A reconsideration of cf. *Psilophyton princeps* (Croft and Lang, 1942), a zosterophyll widespread in the Lower Old Red Sandstone of South Wales. *Bot. J. Linn. Soc.* **100**, 293–318.

Fairon-Demaret, M. (1974) Nouveaux specimens du genre *Leclercqia* Banks, H. P., Bonamo, P. M. et Grierson, J. D. (1972) du Givetian (?) du Queensland (Australia). *Bull. Inst. Roy. Sci. Nat. Belg.* **50** 1–4.

Fairon-Demaret, M. (1978) *Estinnophyton gracile* gen. et sp. nov., a new name for specimens previously determined *Protolepidodendron wahnbachense* Kräusel and Weyland, from Siegenian of Belgium. *Bull. Acad. Belg. (Sci.)* **64**, 597–610.

Fairon-Demaret, M. (1986) Some uppermost Devonian megafloras: a stratigraphic review. *Ann. Soc. Géol. Belg.* **T109–1986** 43–48.

Fanning, U., Richardson, J. B. & Edwards, D. (1988) Cryptic evolution in an early land plant. *Evolutionary trends in plants* **2** 13–24.

Fanning, U., Edwards, D. & Richardson, J. B. (1990) Further evidence for diversity in late Silurian land vegetation. *J. Geol. Soc. Lond.* **147** 725–728.

Fanning, U., Edwards, D. & Richardson, J. B. (in press). A new rhyniophytoid from the late Silurian of the Welsh Borderland. *Neues Jahrb. Geol. Paläont.*

Garratt, M. J. & Rickards, R. B. (1984) Graptolite biostratigraphy of early land plants from Victoria, Australia. *Proc. Yorks. Geol. Soc.* **44** 377–384.

Gensel, P. G. & Andrews, H. N. (1984) *Plant life in the Devonian.* Praeger, New York.

Gothan, W. & Zimmermann, F. (1937) Weiteres über die altoberdevonische Flore von Bögendorf-Liebichau bei Waldenburg. *Jb. Kön. Preuss. Geol. Landesanst. u. Bergakad.* **57** 487–506.

Hao, S.-G. (1989) A new zosterophyll from the Lower Devonian (Siegenian) of Yunnan, China. *Rev. Palaeobot. Palynol.* **57** 155–171.

Heckel, P. H. & Witzke, B. J. (1979) Devonian world palaeogeography determined from distribution of carbonates and related lithic palaeoclimatic indicators. *Special Papers in Palaeontology* **23** 99–123.

Høeg, O. A. (1942) The Downtonian and Devonian Flora of Spitsbergen. *Norges Svalbard-Og Ishavs-Undersøkelser, Skrifter* **83** 1–228.

Hueber, F. M. (1971) Early Devonian land plants from Bathurst Island, District of Franklin. *Pap. Geol. Surv. Can.* **71–28** 1–17.

Hueber, F. M. (1983) A new species of *Baragwanathia* from the Sextant Formation (Emsian), Northern Ontario, Canada. *Bot. J. Linn. Soc.* **86** 57–79.

Jaeger, H. (1962) Das Alter der ältesten bekannten Landpflanzen (*Banagwanathia-*Flora) in Australien auf Grund der begleitenden Graptolithen. *Paläont. Zeitschr.* **36** 7.

Jarvis, E. (1990) New palynological data on the age of the Kiltorkan Flora of County Kilkenny, Ireland. *J. Micropalaeont.* **9** 87–94.

Jeram, A. J., Selden, P. A. & Edwards, D. (1990) Land animals in the Silurian: arachnids and myriapods from Shropshire, England. *Science* **250** 658–661.

Kaplan, L. I. & Senkevich, M. A. (1977) New data on the stratigraphy of Central Kazakhstan. In: *Palaeontology of the Lower Palaeozoic of the Urals.* Sverdlovsk, pp. 92–99.

Kent, D. V. & Van Der Voo, R. (1990) Palaeozoic palaeogeography from palaeomagnetism of the Atlantic-bordering continents. In: McKerrow, W. S. & Scotese, C. R. (eds). *Palaeozoic palaeogeography and biogeography. Geol. Soc. Lond. Mem.* **12** 49–56.

Klitzsch, E., Lejal-Nicol, A. & Massa, D. (1973) Le Siluro-Dévonien a psilophytes et lycophytes du bassin de Mourzouk (Libye). *C. R. Acad. Sci., Paris (Sér. D)* **227** 2465–2467.

Kräusel, R. (1937) Der Verbreitung der Devonfloren. *C. R. 2e Congr. Inst. Strat. Géol. Carbon.* (Heerlen, 1935) **2** 527–537.

Lang, W. H. & Cookson, I. C. (1935) On a flora, including vascular land plants, associated with *Monograptus*, in rocks of Silurian age, from Victoria, Australia. *Phil. Trans. Roy. Soc. Lond. B* **224** 421–449.

Larsen, P.-H., Edwards, D. & Escher, J. C. (1986) Late Silurian plant mega-fossils from the Peary Land Group, North Greenland. *Rapp. Gronlands Geol. Unders* **133** 107–112.

Leclercq, S. (1940) Contribution à l'étude de la flore de Dévonien de Belgique. *Mém. Acad. Roy. Belge. Classe Sci.* **12** 1–65.

Lee H.-H. & Tsai, C.-Y. (1978) *III. Devonian floras of China.* Nanjing (Paper prepared for the International Symposium on the Devonian System 1978).

Lejal-Nicol, A. (1975) Sur une nouvelle à flore Lycophytes du Dévonien inférieur de la Libye. *Palaeontographica B* **151** 52–96.

Lejal-Nicol, A. & Moreau-Benoit, A. (1979) Sur les plantes vasculaires dans le Dévonien de Libye. *Rev. Palaeobot. Palynol.* **27** 193–210.

Lessuise, A. & Fairon-Demaret, M. (1980) Le gisement a plantes de Niaster (Aywaille, Belgique): repere biostratigraphique nouveau aux abords de la limité Couvinien–Givetian. *Ann. Soc. Géol. Belge* **103** 157–181.

Li C.-S. (1990) *Minarodendron cathaysiense* (gen. et comb. nov.) a lycopod from the late Middle Devonian of Yunnan, China. *Palaeontographica B* **220** 97–117.

Li C.-S. & Edwards, D. (in press). A new genus of early land plants with novel strobilar construction from the Lower Devonian Posongchong Formation, Yunnan Province China, *Palaeontology.*

Li C.-S. & Hsü J. (1987) Studies on a New Devonian plant *Protopteridophyton devonicum* assigned to primitive fern from South China. *Palaeontographica B* **207** 111–131.

Matten, L. C. (1974) The Givetian flora from Cairo, New York: *Rhacophyton, Triloboxylon* and *Cladoxylon. Bot. J. Linn. Soc.* **68** 303–318.

Matten, L. C. (1981) *Svalbardia banksii* sp. nov. from the Upper Devonian (Frasnian) of New York State. *Amer. J. Bot.* **68** 1383–1391.

Meyen, S. V. (1982) The Carboniferous and Permian floras of Angaraland (a synthesis). *Biol. Mem.* **7** 1–109.

Meyen, S. V. (1987) *Fundamentals of palaeobotany.* Chapman & Hall, London.

Nikitin, I. F. & Bandaletov, S. M. (1986) *The Tokrau horizons of the Upper Silurian Series Balkhash segment.* Nauka, Alma Ata [In Russian].

Petrosian, N. M. (1968) Stratigraphic importance of the Devonian flora of the USSR. In: Oswald, D. H. (ed.) *International Symposium on the Devonian System, Calgary, 1967.* Alberta Society of Petroleum Geology, Calgary, pp. 579–586.

Raymond, A. (1987) Paleogeographic distribution of Early Devonian plant traits. *Palaios* **2** 113–132.

Raymond, A., Parker, W. C. & Barrett, S. F. (1985) Early Devonian phytogeography In: Tiffney, B. H. (ed.) *Geological factors and the evolution of plants.* Yale University Press, New Haven and London, pp. 129–167.

Rayner, R. J. (1988) Early land plants from South Africa. *Bot. J. Linn. Soc.* **97** 229–237.

Richardson, J. B. (1988) Late Ordovician and early Silurian cryptospores and miospores from Northeast Libya. In: El-Aaranti, A. *et al.*, (eds). *Subsurface palynostratigraphy of north-east Libya,* pp. 89–109.

Richardson, J. B. & Edwards, D. (1989) Sporomorphs and plant megafossils. In: Holland, C. H. & Bassett, M. G. (eds). *A global standard for the Silurian system.* National Museum of Wales (Geological Series Number 9), Cardiff, pp. 216–226.

Richardson, J. B. & McGregor, D. C. (1986) Silurian and Devonian spore zones of the Old Red Sandstone Continent and adjacent regions. *Geol. Surv. Can. Bull.* **364** 1–79.

Richardson, J. B., Rasul, S. M. & Al-Ameri, T. (1981) Acritarchs, microspores and correlation of the Ludlovian–Downtonian and Silurian–Devonian boundaries. *Rev. Palaeobot. Palynol.* **34** 208–224.

Scheckler, S. E. (1975) A fertile axis of *Triloboxylon ashlandicum* a progymnosperm from the Upper Devonian of New York. *Amer. J. Bot.* **62** 923–934.

Scheckler, S. E. (1978) Ontogeny of progymnosperms: II Shoots of Upper Devonian Archaeopteridales. *Can. J. Bot.* **56** 3136–3170.

Scheckler, S. E. (1986a) Geology, floristics and paleoecology of late Devonian coal swamps from Appalachian Laurentia (U.S.A.). *Ann. Soc. Géol. Belge* **109** 209–222.

Scheckler, S. E. (1986b) Old Red Continent facies in the late Devonian and early Carboniferous of Appalachian North America. *Ann. Soc. Géol. Belge* **109** 223–236.

Scheckler, S. E. (1986c) Floras of the Devonian–Mississippian transition. In: Broadhead, T. W. (ed.). *Land plants: notes for a short course,* 15. Department of Geological Sciences Studies in Geology, University of Tennessee, pp. 81–96, i–iv.

Scheckler, S. E., Basinger, J. F. & Hill, S. A. (1990) Floristic evolution in the Devonian Okse Bay Group of Ellesmere, Arctic Canada. *Amer. J. Bot.* **77** (Suppl.) 93.

Schweitzer, H. J. (1983) Die Unterdevonflora des Rheinlandes. *Palaeontographica B* **189** 1–138.

Schweitzer, H.-J. (in press). *Psilophyton burnotense* oder *Psilophyton goldschmidtii* oder *Margophyton goldschmidtii? Cour. Forsch. Inst. Senckenberg, Frankfurt.*

Schweitzer, H. J. & Cai C.-Y. (1987) Beiträge zur Mitteldevon-flora Südchinas. *Palaeontographica B* **207** 1–109.

Scotese, C. R. & McKerrow, W. S. (1990) Revised world maps and introduction. In: McKerrow, W. S. & Scotese, C. R. (eds). *Palaeozoic palaeogeography and biogeography. Geol. Soc. Lond. Mem.* **12** 1–21.

Senkevich, M. A. (1982) Stratigraphic significance of lycopods in the Devonian. *Transactions, Academy of Sciences of the USSR Siberian Branch of Institute of Geology and Geophysics* **483** 84–90 [In Russian].

Stepanov, S. A. (1975) Phytostratigraphy of the key sections of the Devonian of the marginal parts of the Kuznetsk. *Transactions of the Siberian Institute of Geology, Geophysics and Mineral Resources* **211** 1–150 [In Russian].

Stockmans, F. (1940) Végétaux éodévoniens de la Belgique. *Mém. Mus. Roy. d'Hist. Nat. Belge* **93** 1–90.

Streel, M., Higgs, K., Loboziak, S., Riegel, W. & Steemans, P. (1987) Spore stratigraphy and correlation with faunas and floras in the type marine Devonian of the Ardenne–Rhenish Regions. *Rev. Palaeobot. Palynol.* **50** 211–219.

Streel, M., Fairon-Demaret, M. & Loboziak, S. (1990a) Givetian–Frasnian phytogeography of Euramerica and western Gondwana based on miospore distribution. In: McKerrow, W. S. & Scotese, C. R. (eds). *Palaeozoic, palaeogeography and biogeography. Geol. Soc. Lond. Mem.* **12** 291–296.

Streel, M., Fairon-Demaret, M., Gerrienne, P., Loboziak, S. & Steemans, P. (1990b) Lower and Middle Devonian miospore-based stratigraphy in Libya and its relation to the megafloras and faunas. *Rev. Palaeobot. Palynol.* **66** 229–242.

Tanner, W. R. (1983) *A fossil flora from the Beartooth Butte Formation of Northern Wyoming.* Unpublished PhD thesis, Southern Illinois University, Carbondale.

Tims, J. D. & Chambers, T. C. (1984) Rhyniophytina and Trimerophytina from the early land flora of Victoria, Australia. *Palaeontology* **27** 265–279.

Woodrow, D. L., Dennison, J. M., Ettensohn, F. R., Sevon, W. T. & Kirchagasser, W. T. (1988) Middle and Upper Devonian stratigraphy and palaeogeography of the Central and Southern Appalachians and Eastern Midcontinent, U.S.A. In: McMillan, N. J., Embry, A. F. & Glass, D. J. (eds). *Devonian of the World.* Vol. I. Canadian Society of Petroleum Geologists (Memoir 14), Calgary, pp. 277–301.

Yurina, A. L. (1965) A new prefern from the Middle Devonian of Kazakhstan. *Paleont. Zh.* **3** 119–122 [In Russian].

Yurina, A. L. (1988) *The middle and late Devonian floras of northern Eurasia.* Nauka, Moscow and Leningrad [In Russian].

Zakharova, T. V. (1981) On the systematic position of the species '*Psilophyton*' *goldschmidtii* from the Lower Devonian of Eurasia. *Palaeont. Zh.* **23** 111–118 [In Russian].

Ziegler, A. M., Bambach, R. K., Parrish, J. T., Barrett, S. F., Gierlowski, E. H., Parker, W. C., Raymond, A. & Sepkoski, J. J. (1981) Paleozoic biogeography and climatology. In: Niklas, K. J. (ed.). *Paleobotany, paleoecology and evolution, Vol. 2.* Praeger, New York, pp. 231–266.

4

Carboniferous and Permian palaeogeography

C. J. Cleal and B. A. Thomas

The last two periods of the Palaeozoic were a time of marked climatic and geographical change (Ziegler *et al.* 1979, Rowley *et al.* 1985), and of significant diversification in terrestrial vegetation (Niklas *et al.* 1980). These factors together have resulted in a significant provincialism in plant fossil distribution, giving them a considerable value for palaeogeographical work. There have been several attempts to analyse the available data, the most significant being Vakhrameev *et al.* (1978) for Europe and Asia, Read & Mamay (1964), Pfefferkorn & Gillespie (1980) for North America, and Plumstead (1973) and Archangelsky & Arrondo (1975) for the southern continents. There have also been attempts to analyse the data statistically (Rowley *et al.* 1985, Raymond 1985, Raymond *et al.* 1985, Chaloner & Creber 1988). The following discussion will summarize this work, describing the distribution and main palaeobotanical characteristics of the various phytochoria that have been recognized. It deals almost exclusively with the adpression record, as only very few petrifactions from this stratigraphical interval are known outside of palaeoequatorial Laurasia.

Such a discussion needs to be based around viable palaeogeographies for the time. Various reconstructions have been proposed over the last 20 years or so (reviewed by Tarling 1980, 1985, Hallam 1983, Rowley *et al.* 1985). Although there is consensus as to the general configuration of the palaeocontinents, there are differing views as to some of the details. The base maps used here are those developed by Scotese & McKerrow (1990). They are the most recently produced, and seem to agree with most of the palaeophytogeographical evidence presented here. However, there are some discrepancies, such as the relative positions of the Kazakhstania and Angara blocks, and of the north and south China blocks. These discrepancies will be discussed further in sections 4.3.3.2 and 4.5.3.

4.1 PALAEOPHYTOGEOGRAPHICAL MODEL

Meyen's now classic diagram (in Chaloner & Meyen 1973, fig. 3) has provided a generally acceptable model for the development of Carboniferous and Permian phytochoria. From the relatively uniform composition of the Devonian assemblages (see 3.6), there is a gradual increase in provincialism until the top of the Permian, where five discrete palaeokingdoms can be distinguished. Details of this model have been discussed extensively (e.g. Chaloner & Lacey 1973, Raymond 1985, Li & Yao 1985, Meyen 1987, Allen & Dineley 1988, Chaloner & Creber 1988) and it now appears that certain aspects need to be revised. A revised version of Meyen's model is therefore provided in Fig. 4.1, the evidence for which will be discussed through this chapter. The palaeogeographic distribution of the phytochoria is shown in Figs 4.2, 4.4 and 4.6.

It is impossible in the space available to provide comprehensive details as to how the palaeokingdoms and palaeoareas can be recognized. As a general guide, however, diagrammatic representations of the distribution of some of the commoner form-genera are provided in Figs. 4.3, 4.5 and 4.7. The charts have been divided chronostratigraphically into Lower Carboniferous, Upper Carboniferous and Permian (see section 1.6 for a discussion of the usage of Lower and Upper Carboniferous adopted in this book). These chronostratigraphic units represent fairly 'natural' divisions, the boundaries between which often reflect quite significant changes in the lowland floras. According to the radiometric time-scale of Harland *et al.* (1982), they also represent approximately uniform time intervals (*c.*35–40 million years), although some of the evidence has been queried by Leeder (1988).

4.2 EURAMERIA PALAEOKINGDOM

This phytochorion has also been referred to as the Amerosinian (Pfefferkorn & Gillespie 1980), Equatorial (Raymond *et al.* 1985), Circum-Tethyan (Allen & Dineley 1988) and Euramerican (Chaloner & Creber 1988), but the original name used by Chaloner & Meyen (1973) has been retained here. It represents some of the best documented assemblages from the Carboniferous and Permian, as reported from Europe and North America.

4.2.1 Distribution

In the Lower Carboniferous, Euramerian-type assemblages have an almost global distribution, and this partly explains Jongmans's (1952) view about the geographical uniformity of floras of this age. They occur throughout the Laurasia, Kazakhstania and Cathaysia palaeocontinents (Vakhrameev *et al.* 1978), and also probably into the southern palaeolatitudes of Gondwana. The affinities of the Lower Carboniferous assemblages of Gondwana have been the subject of some disagreement, and they have sometimes been assigned to their own palaeokingdom (e.g. Raymond *et al.* 1985). However, this has been largely based on the *Botrychiopsis–Nothorhacopteris* assemblages of Gondwana, whose stratigraphic position is now recognized as Upper

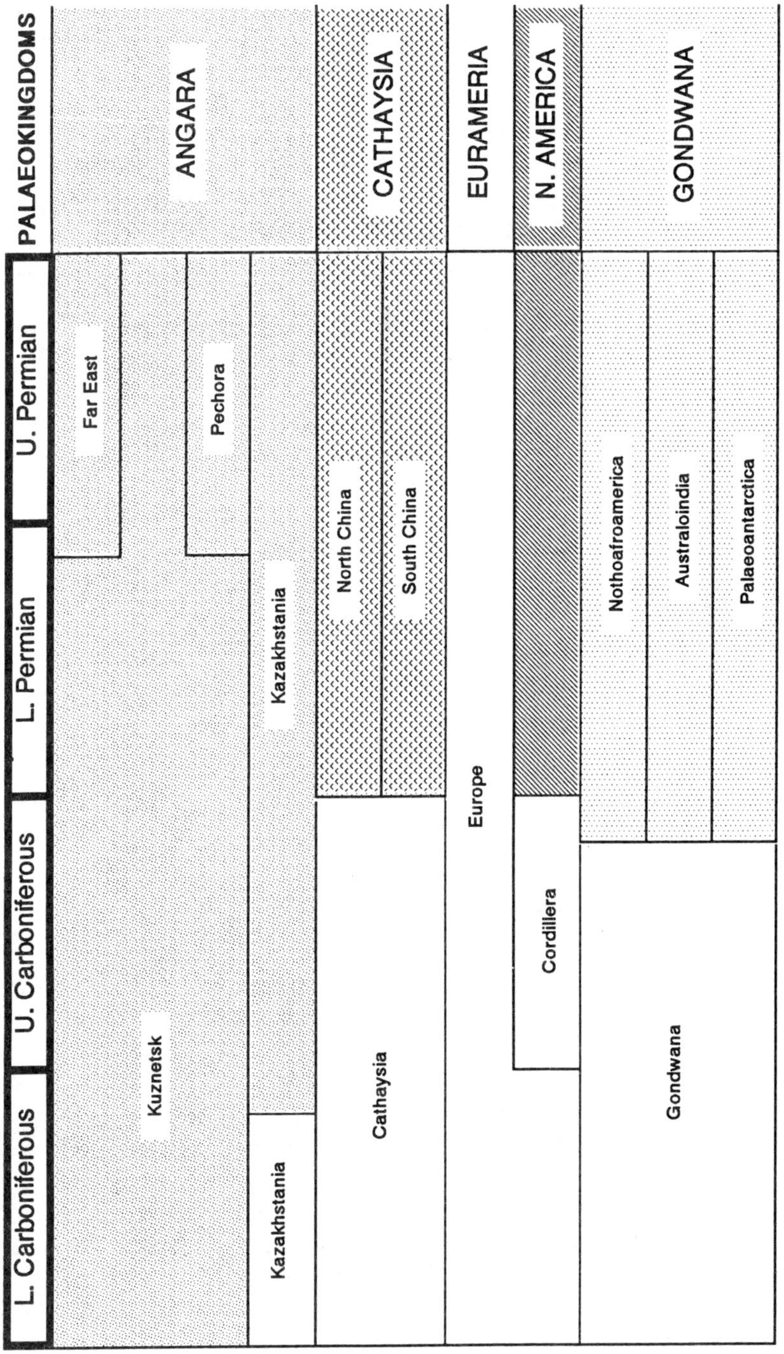

Fig. 4.1 — The evolution of Carboniferous and Permian palaeophytogeography. Shading patterns and names in capitals represent different palaeokingdoms; names in lower case represent palaeoareas.

Carboniferous (Wagner *et al.* 1985). The Gondwana assemblages now thought to be Lower Carboniferous (as reviewed by Wagner *et al.* 1985) are of very restricted composition, but contain few form-genera which do not occur in the rest of the Eurameria Palaeokingdom.

In the Upper Carboniferous, the Eurameria Palaeokingdom becomes limited to equatorial palaeolatitudes in Laurasia, northern Gondwana and Cathaysia, the Gondwana assemblages being separated into their own palaeokingdom. It nevertheless extends in a wide band from present-day western North America to the Caucasus and southern Kazakhstan (Vakhrameev *et al.* 1978). In the Lower Permian, it becomes further restricted, partly due to the separation of the western North America and Cathaysia assemblages into their own palaeokingdoms (see sections 4.5 and 4.6). The only known Upper Permian Eurameria assemblages are the so-called Zechstein 'flora' of Europe (Schweitzer 1986).

4.2.2 Floristic history

Particularly in southern Laurasia and northern Gondwana, the Early Carboniferous lowlands appear to have been covered by large areas of forest, dominated by arborescent lycophytes (Lepidocarpales), with the characteristic *Stigmaria* rooting structures and *Flemingites* (*Lepidostrobus auct.*) and *Lepidocarpon* cones. Other significant elements in these forests were sphenophytes (*Archaeocalamitaceae* Bowmanitales), progymnosperms and pteridosperms (mainly Calamopityales and Lagenostomales), although the detailed ecological relationships between these plants remain somewhat unclear.

The mid-Carboniferous boundary marks a significant change in the palaeoequatorial floras, probably triggered by the onset of major palaeoantarctic glaciation (Wagner 1982, Pfefferkorn & Gillespie 1982). The lowland forests became considerably more prolific and produced thick peat deposits (subsequently forming coal seams). In the early Late Carboniferous, arborescent lycophytes remained dominant (Phillips 1980), but towards the end of the Carboniferous, climatic changes resulted in them being largely replaced by medullosan pteridosperms, marattialean ferns and cordaites (Phillips 1980, Phillips & Peppers 1984). Throughout the Late Carboniferous, there were also stands of pteridosperms (Lagenostomales, Trigonocarpales and Callistophytales), ferns (e.g. Marattiales, Zygopteridales, Botryopteridales) and sphenophytes (Calamostachyales, *Bowmanitaceae*) probably growing on raised levee-banks (Gastaldo 1987). Although only a relatively small part of the original forest community, the remains of the levee-bank flora usually dominate the adpression assemblages and include most of the distinctive elements of the Eurameria Palaeokingdom. There is some evidence that the hinterlands to these forests contained conifer and cordaite forests (Scott & Chaloner 1983, Lyons & Darrah 1989), but their remains rarely find their way into the fossil record (see also Leary 1974, 1981, for discussions on extra-basinal floras in the Late Carboniferous).

At about the start of the Permian, the palaeoequatorial climate became significantly more arid and the dense forests largely disappeared. Conifers (*Walchiaceae*, *Voltziaceae*) formed a major component of the lowland floras for the first time, although rarely if ever forming dense forests. Other significant elements appear to have been marattialean ferns and pteridosperms (Peltaspermales, Callistophytales).

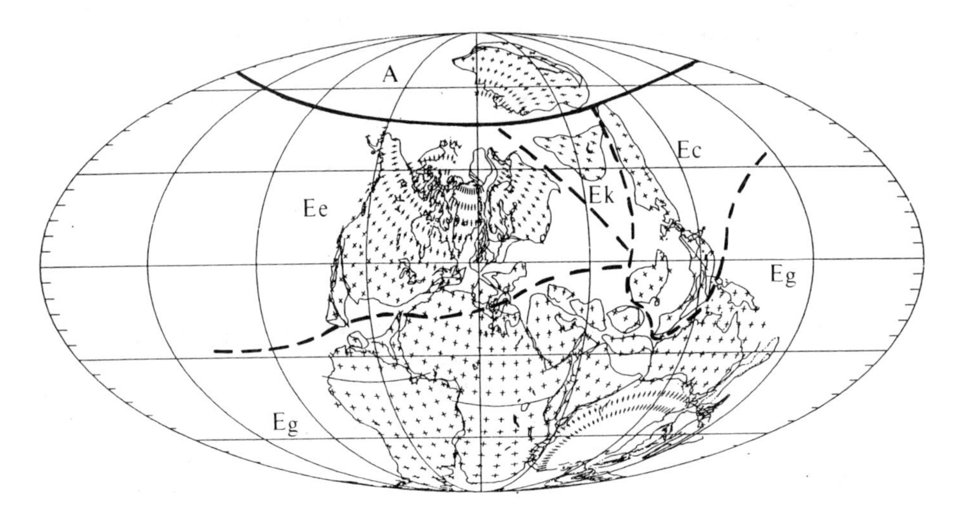

Fig. 4.2 — Distribution of Lower Carboniferous phytochoria. Base map adapted from Scotese & McKerrow (1990). A=Angara Palaeokingdom; Ee=Europe Palaeoarea; Ek=Kazakhstania Palaeoarea; Ec=Cathaysia Palaeoarea; Eg=Gondwana Palaeoarea.

In the later part of the Permian, climatic aridification caused the palaeoequatorial vegetation to become significantly sparser, with only occasional stands of mainly conifers, peltasperm pteridosperms and cycad-like plants.

4.2.3 Palaeoareas

There have been several attempts to subdivide the Eurameria Palaeokingdom, by recognizing either particular endemic taxa (Gothan 1951, Remy & Remy 1959, Pfefferkorn & Gillespie 1980) or peculiarities in the general taxonomic composition of the assemblages (Rowley *et al.* 1985, Raymond 1985, Raymond *et al.* 1985). The authors have recognized the following palaeoareas, based on their distinctive composition.

4.2.3.1 Europe

This is in many ways the core of the Eurameria Palaeokingdom, representing the most diverse and well studied assemblages (Vakhrameev *et al.* 1978, Pfefferkorn & Gillespie 1980 and Wagner 1985 give general reviews). In the Lower Carboniferous it includes the assemblages from Europe and North America (Read 1955, Scott 1985), whilst in the Upper Carboniferous it includes the classic 'Coal Measures floras'. In the Permian, evidence becomes sparse, but includes assemblages from the red-bed and 'Zechstein' deposits of Europe (Doubinger 1956, Kerp & Fichter 1985, Schweitzer 1986) and North America (Read & Mamay 1964).

There have been a number of attempts to further subdivide the palaeoarea (e.g. Gothan 1951, Remy & Remy 1959, Pfefferkorn & Gillespie 1980, Raymond *et al.* 1985, Rowley *et al.* 1985, Raymond 1985, Allen & Dineley 1988). However, this has been based mainly on the recognition of subtle variations in form-genus composition, or on endemism at the rank of species. At the rank of form-genus there is, in

Palaeokingdom	Euramerica				Angara
Palaeoarea	Gondwana	Europe	Cathaysia	Kazakhstania*	Kuznetsk
Archaeocalamites	1 2	1 2 3	1 2 3	1 2 3	2 3
Cardiopteridium	2	2 3	2	1 2 3	3
Lepidodendropsis	1 2	1 2	1	1	1
Lepidodendron	2	1 2 3	1 2	1 2 3	
Adiantites	2	1 2 3	1 2	3	1
Sphenopteridium	2	1 2 3	1 2	2 3	
Sublepidodendron	1 2	1	1 2	2 3	
Rhodeopteridium	2	1 2 3	1 2		
Diplothmema	2	1 2 3			
Eleutherophyllum	2	2 3			
Leptophloeum	1				
Archaeosigillaria	2				
Cyclostigma	2				
Furqueia	2				
Eskdalia		2			1 2
Neuropteris		2 3	2	2 3	3
Mesocalamites		2 3	2 3	2 3	3

* The Kazakhstania Palaeoarea becomes part of the Angara Palaeokingdom towards the top of the Lower Carboniferous.

1 - Tournaisian and basal Visean; 2 - most of Visean; 3 - top Visean and basal Namurian.

Palaeokingdom	Euramerica				Angara
Palaeoarea	Gondwana	Europe	Cathaysia	Kazakhstania*	Kuznetsk
Sphenophyllum		1 2 3	2 3	1 2 3	1 2 3
Calamites		3			
Rhacopteris		1 2 3	1		
Triphyllopteris		1 2	1 2		
Lyginopteris		2 3	2 3		
Aneimites		1 2			
Diplopteridium		2			
Presigillaria		2			
Psygmophyllum		2			
Chwydia		2			
Alloiopteris		2			
Spathulopteris		2 3			
Fryopsis		2 3			
Lepidophloios		2 3			
Archaeopteridium		3			
Eolepidodendron			1	2	
Paripteris			2 3		
Eusphenopteris			3		

* The Kazakhstania Palaeoarea becomes part of the Angara Palaeokingdom towards the top of the Lower Carboniferous.

1 - Tournaisian and basal Visean; 2 - most of Visean; 3 - top Visean and basal Namurian.

Fig. 4.3 — Occurrence of key form-genera in Lower Carboniferous palaeoareas. See text for main sources of data.

Palaeokingdom	Eurameria				Angara
Palaeoarea	Gondwana	Europe	Cathaysia	Kazakhstania*	Kuznetsk
Karinopteris			3		
Angaropteridium				2 3	3
Koretrophyllites				3	3
Neubergia				1	
Porodendron				1	
Dzungarodendron				1	
Eolepidophloios				1	
Neurocardiopteris				1	
Cardioneura				1 2 3	
Caenodendron				1 2 3	
Hartungia					1
Pseudolepidodendron					1
Ursodendron					1 2
Tomiodendron					1 2 3
Angarophloios					2
Lophiodendron					2 3
Angarodendron					2 3
Abacodendron					3
Chacassopteris					3
Paracalamites					3
Siberiodendron					3

* The Kazakhstania Palaeoarea becomes part of the Angara Palaeokingdom towards the top of the Lower Carboniferous.

1 - Tournaisian and basal Visean; 2 - most of Visean; 3 - top Visean and basal Namurian.

Fig. 4.3 — Occurrence of key form-genera in Lower Carboniferous palaeoareas. See text for main sources of data.

fact, a remarkable uniformity within the European assemblages, even as far east as the Caucasus and southern Kazakhstan (Vakhrameev *et al.* 1978). It is also of interest that the Upper Carboniferous coal ball assemblages are essentially uniform in composition between the Ukraine and Oklahoma (Phillips & Peppers 1984). In view of the subtle distinctions between these phytochoria, and thus the difficulty in recognizing them in practice, they are not shown in Figs 4.2, 4.4 and 4.6.

4.2.3.2 Gondwana

This is only recognized in the Lower Carboniferous; at higher stratigraphic levels it develops into a discrete palaeokingdom (see section 4.4). The best documented assemblages are from Argentina and Australia (Wagner *et al.* 1985). From the evidence summarized in Fig. 4.3 (see also section 4.2.1 above) the palaeoarea is characterized by remains of non-arborescent lycophytes, including the typically Devonian form *Leptophloeum*. There are also apparently some endemic form-genera, such as *Furqueia* and *Charnelia*. It is perhaps worth repeating here for emphasis that some Gondwana assemblages traditionally placed in the Lower Carboniferous (the so-called *Botrychiopsis–Nothorhacopteris* assemblages) are now known to have originated from the Upper Carboniferous.

4.2.3.3 Kazakhstania

This is part of the Eurameria Palaeokingdom between the Tournaisian and Upper Visean, above which it becomes part of the Angara Palaeokingdom (see section

4.3.3.2). Its isolated palaeogeographic position has resulted in assemblages of a highly distinctive character, with some 45% of the form-genera being endemic (data taken from Vakhrameev *et al.* 1978, Raymond 1985 and Raymond *et al.* 1985). Form-genera useful for distinguishing it from the Europe Palaeoarea include *Neubergia*, *Porodendron*, *Dzungarodendron*, *Caenodendron* and *Cardioneura*. The palaeoarea is best known from a belt of depositional basins in Central-Southern Kazakhstan (*sensu* Abdulin *et al.* 1975), between lakes Tengiz and Zaysan; the best known is probably the Karaganda Basin. The evidence is usefully surveyed by Radchenko (1967, 1985), Vakhrameev *et al.* (1978) and Meyen (1987).

4.2.3.4 Cathaysia

Lower Carboniferous and lower Upper Carboniferous assemblages from the Cathaysia palaeocontinent (at least, those not belonging to the Angara and Gondwana palaeokingdoms; Li & Yao 1985) are broadly comparable with those of the Europe Palaeoarea. The only noteworthy difference is that some form-genera occur stratigraphically lower in Cathaysia than elsewhere in the palaeokingdom (Laveine *et al.* 1987). The upper Upper Carboniferous assemblages take on a more distinctive character, however, with about 30% of the Cathaysian form-genera being endemic (based on data from Vakhrameev *et al.* 1978, Wagner *et al.* 1983, Zhao & Wu 1985), including *Lobatannularia, Cathaysiodendron, Emplectopteris, Emplectopteridium* and *Dictyocallipteridium*. This is probably because the climatic aridification which had started at this time in Laurasia had much less effect in Cathaysia. It marks the start of a gradual palaeophytogeographic separation of the two palaeocontinents, which, in the Permian, results in the development of discrete palaeokingdoms (section 4.5).

Topmost Carboniferous assemblages of the Cathaysia Palaeoarea occur mainly on the North China Block, but have also been reported from Sumatra (Wagner *et al.* 1983, Li & Yao 1985). There has been some debate as to the timing of the accretion of the North and South China blocks, but it has been recently argued that it was as early as the Devonian (e.g. Laveine *et al.* 1987).

4.2.3.5 Cordillera

The concept of a separate phytochorion for Upper Carboniferous assemblages from western North America (Wyoming, Utah, Nevada) was put forward by Read (*in* Read & Mamay 1964). The argument was based mainly on the presence of conifers and a paucity of lycophytes in many of the assemblages (e.g. Arnold 1941). The documentation of the assemblages from this area is still relatively poor, one of the few to be studied in recent years being from the Manning Shale of Utah (Tidwell 1967). However, the existence of this palaeoarea in the Upper Carboniferous is given additional credence by the development of a distinct palaeokingdom in the same region in the Permian (section 4.6).

4.2.3.6 Oregon

This phytochorion was established by Pfefferkorn & Gillespie (1980) based on a single assemblage from Oregon containing *Phyllotheca* and *Dicranophyllum*. Clearly more evidence is needed to confirm its palaeophytogeographic position, but there is some evidence that this part of North America may have been palaeogeographically isolated from the rest of Laurasia.

4.3 ANGARA PALAEOKINGDOM

Assemblages originating from northern palaeolatitudes are distinct from the Eurameria Palaeokingdom throughout the Carboniferous and Permian, and probably reflect provincialism already developed in the Devonian (see section 3.6). By the Upper Carboniferous, 'typical' Angaran assemblages (as represented by the Kuznetsk Palaeoarea; see section 4.3.3.1) have virtually no form-genera in common with the other palaeokingdoms. The evidence has been reviewed by Jongmans (1939), Vakhrameev *et al.* (1978) and Meyen (1982, 1987).

4.3.1 Distribution

In the basal Carboniferous, this palaeokingdom is restricted to the Siberia palaeocontinent. Towards the end of the Early Carboniferous, however, Kazakhstania collided with this palaeocontinent (Zhang *et al.* 1984), and assemblages from the former region take on the general aspect of the Angara Palaeokingdom. By the end of the Carboniferous or the start of the Permian, the Angara palaeocontinent came into contact with Laurasia (Tarling 1980) and Cathaysia, and so it is not surprising to find Angara-type assemblages in the Permian of northern Laurasia (e.g. Wagner *et al.* 1982) and north China (Durante 1983).

4.3.2 Floristic history

Meyen (1982, 1987) reviews Angaran floristics and provides more detail than is possible here. As in the equatorial palaeolatitudes, the Early Carboniferous floras were dominated by lycophytes. Unlike the equatorial forms, however, they were relatively small, lacked persistent leaf-scars on the stems, and did not have *Stigmaria*-like rooting structures. Also, the fructifications were not in the form of discrete cones, but the sporangia were attached directly to vegetative axes. Pteridosperms and progymnosperms were rare, but there is evidence of primitive ferns, such as *Chacassopteris*. The distinctive nature of the Angaran floras of this time has been attributed to climatic aridity in the northern palaeolatitudes (Meyen 1987).

The collision between the Siberia and Kazakhstania palaeocontinents towards the end of the Early Carboniferous caused the Angara Palaeokingdom to cover a much larger area, and its floras to become more diverse. In particular, the number of pteridosperms and sphenophytes seems to have increased at this time.

As in the equatorial palaeolatitudes, the Angaran floras seem to have become significantly more prolific at the mid-Carboniferous boundary, as shown by the development of thick peat deposits. There was also a significant increase in floral diversity in the early Late Carboniferous, and a change from lycophyte-dominated to pteridosperm- and then cordaite-dominated forests. The cordaites continued to dominate the lowlands in the Permian, but conifers, sphenophytes, pteridosperms and ferns also occur locally. Under certain conditions, abundant bryophyte floras developed, producing some of the oldest known assemblages to be dominated by these plants.

4.3.3 Palaeoareas

From the upper Lower Carboniferous to the Lower Permian two palaeoareas can be recognized in the Angara Palaeokingdom (Kuznetsk and Kazakhstania), which probably reflects the origin of this palaeocontinent as two separate plates (section 4.3.1). In the Upper Permian, two further palaeoareas develop: Pechora and Far East.

4.3.3.1 Kuznetsk

This represents the best developed Angaran assemblages, as found in the centre of the Siberia palaeocontinent. It probably reflects the core from which the Angaran floras spread out to remoter parts of the palaeocontinent (Neuburg 1961). The best documented assemblages are from the Minusinsk and Kuznetsk basins near Novisibirsk, southern Siberia (Neuburg 1948, Gorelova *et al.* 1973). However, there are also numerous other recorded occurrences in central and eastern Siberia, and are reviewed by Vakhrameev *et al.* (1978) and Meyen (1982, 1987). They share few form-genera with the other palaeokingdoms, and include common endemic form-genera such as *Lophiodendron, Angarodendron, Abacodendron, Chacassopteris, Paracalamites, Siberiodendron, Tomiodendron, Angarophloios, Tchirkoviella, Angaridium, Prynadaeopteris* and *Dichophyllites*.

4.3.3.2 Kazakhstania

The Upper Carboniferous palaeobotany of this palaeoarea has been well documented by Oshurkova (1967) and Radchenko (1967, 1985) and, as in the Lower Carboniferous, is best known from the Karaganda Basin of central Kazakhstan. It becomes part of the Angara Palaeokingdom in the upper Lower Carboniferous, and has been interpreted as an ecotone between the 'typical' Angaran assemblages of the Kuznetsk Palaeoarea and the Eurameria Palaeokingdom (Meyen 1987). In the lower Upper Carboniferous, Vakhrameev *et al.* (1978) claim that 52% of the Middle Carboniferous Kazakhstania species are Euramerian, whilst 13% are Angaran (an analysis of the data at the rank of form-genus results in essentially similar results — 52% Euramerian, 21% Angaran). This tends to support Allen & Dineley (1988), who include the Kazakhstania assemblages in the Eurameria Palaeokingdom until the end of the Carboniferous. However, the Kazakhstania palaeocontinent had clearly accreted to the Siberia plate by the Late Carboniferous (section 4.3.1), and so assemblages of this age take on an increasingly Angaran aspect, especially towards the top of the Upper Carboniferous. For simplicity, therefore, the authors have opted to include it in the Angara Palaeokingdom throughout the Upper Carboniferous, distinguishing it as a palaeoarea. Endemic Carboniferous form-genera include *Caenodendron, Cardioneura, Neurocardiopteris, Zamiopteris* and *Gaussia*.

In the Permian, the Kazakhstania Palaeoarea (the Subangara 'Area' of Meyen 1987) remains a separate phytochorion. Vakhrameev *et al.* (1978) referred to it as the Ural–Kazakhstan and East-European phytochoria in the lower and upper Permian, respectively, owing to changes in its geographical range and palaeobotanical character. However, as there is an essential continuity with the Kazakhstania Palaeoarea of the Carboniferous, there seems little real reason for changing its name in the Permian. Meyen (1987) has argued that it still seems to be essentially an ecotone between the more typical Angaran assemblages and the Eurameria Palaeokingdom, since it includes a number of elements also found in Laurasia, such as *Lebachia, Ernestiodendron, Ullmannia, Pachypteris* and *Odontopteris*, together with the more typical Angaran form-genera. Particularly in the Upper Permian, however, the plant fossils have become quite distinct such as the *Tatarina* 'flora' found in the upper Tatarian (Gomankov & Meyen 1986) and the Nanshan 'flora' from north-west China (Durante 1983). This is perhaps a justification for raising the Kazakhstania Palaeoarea to the rank of palaeokingdom. However, the traditional view has been followed here and it has been retained as a palaeoarea within the Angara Palaeokingdom (it may be significant that, by the Lower Triassic, the distinction between Kuznetsk and Kazakhstania palaeoareas has essentially disappeared).

Permian assemblages from Uzbekistan and the southernmost parts of Kazakhstan ('Middle Asia') present a number of problems (Sixtel 1975; Sixtel & Savitskaya 1985). They are on the southern edge of the Kazakhstania palaeocontinent, and contain a mixture of Angaran, Cathaysian and reputedly Euramerian taxa. Soviet palaeontologists have mostly classified them as Upper Carboniferous, but the Cathaysian elements clearly point to the Lower Permian (possibly the *Cathaysiopteris whitei* Biozone — see section 5.3). If they are Lower Permian, then the reputed Euramerian affinities of the assemblages become less convincing; all of the taxa claimed to link it with that palaeokingdom are also found in Cathaysian Lower Permian assemblages. A simpler explanation is thus that the Middle Asia assemblages represent a mixture of elements of the North China– Kazakhstania palaeoareas, and indicate that the Cathaysia and Angara palaeocontinents were already near each other, if not in actual contact, by the Early Permian.

4.3.3.3 Pechora

Upper Permian assemblages from near the Pechora Inlet on the north coast of Russia are intermediate in character and geographical position between those typical of the Kuznetsk and Kazakhstania palaeoareas, and have been assigned to a separate Pechora Palaeoarea (Meyen 1983). They are characterized by numerous endemic bryophytes (Fefilova 1978), as well as having a greater abundance of ferns, and the presence of Euramerian elements such as *Oligocarpia* (Meyen 1987). This palaeoarea also appears to be present in the Upper Permian of northern Laurasia (Wagner *et al*. 1982).

4.3.3.4 Far East

This represents assemblages found on the north-eastern margins of the Siberia palaeocontinent. They are generally of more restricted composition than those of the other Angaran palaeoareas, which may reflect floras growing in rather higher palaeolatitudes. Some of the best documented examples have been found in present-day Mongolia (Durante 1976). They are characterized by the absence of callipterids (*sic*) and of *Psygmophyllum*, and the less abundant ferns (Meyen 1987).

4.4 GONDWANA PALAEOKINGDOM

In the lower Lower Carboniferous, the assemblages of the southern palaeolatitudes are broadly similar to those of the equatorial palaeolatitudes, albeit not so diverse (section 4.2.3.2). There is then little palaeobotanic evidence until the middle Upper Carboniferous, when the assemblages develop a character quite distinct from those of the equatorial palaeolatitudes, with there being few form-genera in common. There is in fact a greater superficial similarity with the assemblages of the Angara Palaeokingdom, but this is now recognized to be a result of parallel evolution in different groups of plants adapting to comparable conditions (Meyen 1969). Details of the early evolution of the Gondwana Palaeokingdom are as yet unclear, owing to the limited data available from the middle part of the Carboniferous, but it may be related to the development of palaeoantarctic glaciation. Details of the Gondwana Palaeokingdom assemblages are discussed by Plumstead (1973), Schopf & Askin (1980), Wagner *et al.* (1985) and Archangelsky (1990).

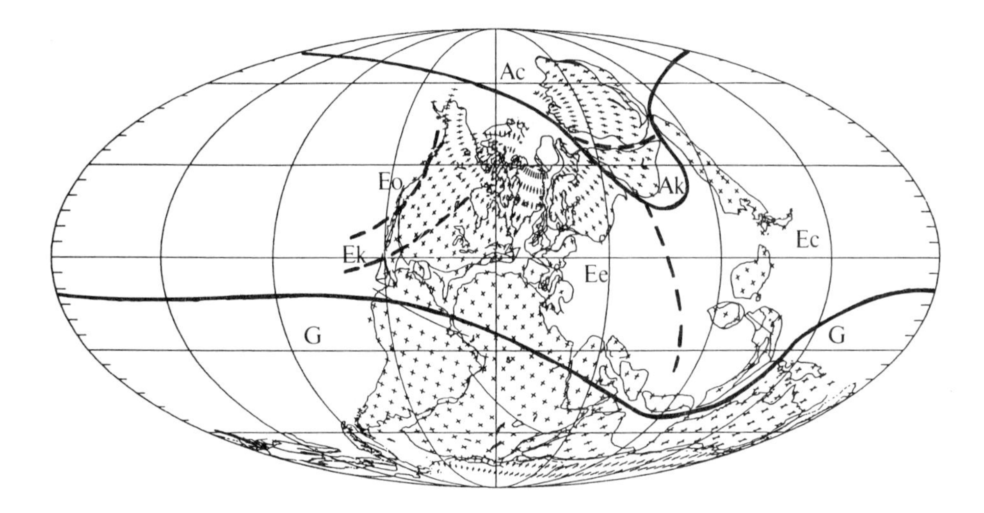

Fig. 4.4 — Distribution of Upper Carboniferous phytochoria. Base map adapted from Scotese & McKerrow (1990). Ac=Kuznetsk Palaeoarea; Ak=Kazakhstania Palaeoarea; Eo=Oregon Palaeoarea; Ek=Cordillera Palaeoarea; Ee=Europe Palaeoarea; Ec=Cathaysia Palaeoarea; G=Gondwana Palaeokingdom.

4.4.1 Distribution

Evidence of the palaeokingdom in the Upper Carboniferous occurs mainly in the eastern and western extremities of the Gondwana palaeocontinent (present-day South America and Australia; Wagner *et al.* 1985). In the Permian, however, it occurs throughout the palaeocontinent, extending north into present-day Turkey, where there is evidence of an ecotone with the Cathaysia Palaeokingdom (e.g. Archangelsky & Wagner 1983).

Palaeokingdom	Gondwana	Eurameria			Angara	
Palaeoarea		Europe	Cordillera	Cathaysia	Kazakhstania	Kuznetsk
Cordaites	2 3	1 2 3	1	1 2 3	3	2 3
Rhodeopteridium	1 2	1	1	1		1
Diplothmema	1 2	1 2	1			
Ginkgophyllum	2 3				3	2 3
Lepidodendropsis	1 2					
Nothorhacopteris	1 2 3					
Botrychiopsis	2 3					
Bergiopteris	2 3					
Bumbodendron	2 3					
Paulophyton	2 3					
Fedekurtzia	2 3					
Neuropteris		1 2 3	1	1 2 3	1 2	1 2 3
Lepidodendron		1 2 3	1	1 2 3	1 2	
Calamites		1 2 3	1	2 3	1 2 3	
Alloiopteris		1 2 3	1		1 2	
Mesocalamites		1 2	1	1	1 2	
Adiantites		1	1		1	
Paripteris		1 2	1	1 2		
Eusphenopteris		1 2 3	1	1 2 3		

1 - most of Namurian; 2 -Westphalian; 3 - Stephanian

Palaeokingdom	Gondwana	Eurameria			Angara	
Palaeoarea		Europe	Cordillera	Cathaysia	Kazakhstania	Kuznetsk
Mariopteris		1 2 3	1	1		
Sigillaria		1 2 3	1	1 2 3		
Asterophyllites		1 2 3	1	1 2 3		
Lepidophloios		1 2 3	1	2		
Sphenophyllum		1 2 3		1 2 3	1 2	2 3
Annularia		1 2 3		3		3
Palmatopteris		1 2 3			1 2	
Renaultia		1 2 3			1 2	
Lyginopteris		1 2			1 2	
Alethopteris		1 2 3		3		
Karinopteris		1 2		1		
Linopteris		2 3		2		
Cyathocarpus		2 3		3		
Odontopteris		2 3		3		
Callipteridium		2 3		3		
Nemejcopteris		3		3		
Neuralethopteris		1 2				
Senftenbergia		1 2 3				
Zeilleria		1 2 3				

1 - most of Namurian; 2 -Westphalian; 3 - Stephanian

Fig. 4.5 — Occurrence of key form-genera in Upper Carboniferous palaeoareas. See text for main sources of data (*continued next page*).

Palaeokingdom	Gondwana	Eurameria			Angara	
Palaeoarea		Europe	Cordillera	Cathaysia	Kazakhstania	Kuznetsk
Oligocarpia		2 3				
Lobatopteris		2 3				
Ptychocarpus		2 3		3		
Asolanus		2 3				
Dicksonites		2 3				
Polymorphopteris		2 3		3		
Eremopteris		2 3				
Reticulopteris		2 3				
Lescuropteris		3				
Megalopteris		2				
Eleutherophyllum		1				
Cardiopteridium		1				
Pseudoadiantites		1 2				
Discopteris		2				
Lonchopteris		2				
Margaritopteris		2				
Palaeoweichselia		2				
Crossopteris			1			
Tingia			1	2 3		

1 - most of Namurian; 2 -Westphalian; 3 - Stephanian

Palaeokingdom	Gondwana	Eurameria			Angara	
Palaeoarea		Europe	Cordillera	Cathaysia	Kazakhstania	Kuznetsk
Lobatannularia				3		
Cathaysiodendron				3		
Emplectopteris				3		
Emplectopteridium				3		
Dictyocallipteridium				3		
Caenodendron					1 2	1
Angaropteridium					1 2 3	1 2 3
Rufloria					3	1 2 3
Phyllotheca					3	2
Paragondwanidium					3	2 3
Neurocardiopteris					1	
Cardioneura					1 2	
Zamiopteris					3	
Gaussia					3	
Angarodendron						1 2
Tomiodendron						1
Paracalamites						2
Angaridium						1 2 3
Prynadaeopteris						3
Angarophloios						3

1 - most of Namurian; 2 -Westphalian; 3 - Stephanian

Fig. 4.5 (*continued*) — Occurrence of key form-genera in Upper Carboniferous palaeoareas. See text for main sources of data.

4.4.2 Floristic history

In contrast to the extensive forests that dominated the equatorial palaeolatitudes, there appear to have been rather more open floras, dominated, for most of the Late Carboniferous, by shrubby protolepidodendrids, sphenophytes and progymnosperms. Retallack (1980) has called this 'the *Botrychiopsis*-tundra', although the old term 'pre-*Glossopteris* flora' is still used by some. The resulting assemblages have a 'primitive' aspect when compared with the Eurameria Palaeokingdom, and have often been regarded erroneously as Lower Carboniferous (e.g. Raymond 1985).

In the very late Carboniferous (or possibly Early Permian), extensive forests developed for the first time in the southern palaeolatitude lowlands. These were dominated mainly by arberialeans (glossopterids *auct.*), whose leaves (e.g. *Glossopteris, Gangamopteris*) form the characteristic components of the assemblages. Also present were some conifers (e.g. *Buradiaceae*), sphenophytes (*Gondwanostachyaceae*) and ferns (Osmundales). In the more northern parts of Gondwana, marattialean ferns and lycophytes occurred.

4.4.3 Palaeoareas

Assemblages from the Upper Carboniferous of South America appear to be more diverse than those from Australia, particularly in lycophytes, but this may simply be because they have been more intensively studied. In the Permian, however, Archangelsky (1990) argues that three phytochoria can be recognized.

4.4.3.1 Nothoafroamerica

This represents assemblages from areas nearer to where the Eurameria Palaeokingdom is found, in present-day South America (e.g. the Paraná Basin of Brazil and the Paganzo Basin of Argentina) and southern Africa (e.g. the Karroo Basin). They reflect temperate, mainly frost-free climates. In addition to the typical Gondwanan Arberiales, there are abundant lycophytes such as *Cyclodendron* (but without the *Stigmaria*-type rooting structures found in the Eurameria Palaeokingdom), Calamostachyales-like sphenophytes and marattialean ferns (*Asterotheca*) (Archangelsky & Arrondo 1969, 1975, Plumstead 1967).

4.4.3.2 Australoindia

The assemblages from the east of Gondwana, in present-day India and Australia), have few lycophytes (although they are represented in the palynological record; Meyen 1987) or the other Eurameria-like elements found in the Nothoafroamerica Palaeoarea (Surange 1975, Retallack 1980). Although also representing plants growing in temperate climates, they were palaeogeographically more isolated than those of the Nothoafroamerica Palaeoarea.

4.4.3.3 Palaeoantarctica

Assemblages from the southern palaeoantarctic consist almost exclusively of arberialeans, possibly reflecting the extreme climatic conditions prevailing (Rigby & Schopf 1969, Gee 1989, Archangelsky 1990).

4.5 CATHAYSIA PALAEOKINGDOM

Unlike most of the palaeoequatorial parts of Laurasia, the Cathaysia palaeocontinent did not suffer significant climatic aridity until the end of the Permian. Consequently, there is not the same type of dramatic difference in plant fossils between the Carboniferous and Permian. The distinctiveness of the Cathaysian assemblages is already evident in the Upper Carboniferous, where they are assigned their own palaeoarea (section 4.2.3.4), but by the Permian they share so few taxa with the typical Eurameria assemblages (i.e. of the Europe Palaeoarea; section 4.2.3.1) that they are assigned to a separate palaeokingdom. They are reviewed by Gu & Zhi (1974), Vakhrameev *et al.* (1978), Asama (1984), Wang (1985) and Li & Yao (1985).

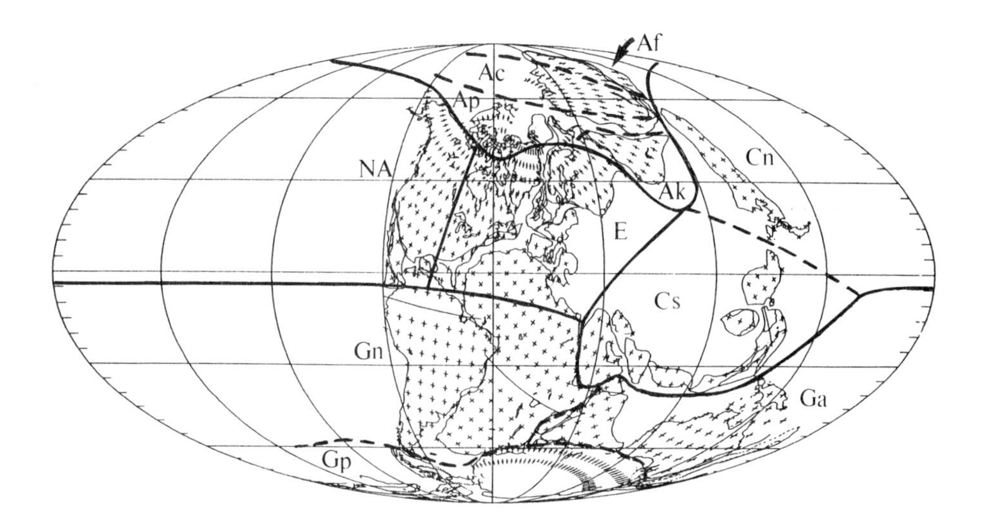

Fig. 4.6 — Distribution of Permian phytochoria. Base map based on Scotese & McKerrow (1990). Af=Far East Palaeoarea; Ac=Kuznetsk Palaeoarea; Ap=Pechora Palaeoarea; Ak=Kazakhstania Palaeoarea; E=Eurameria Palaeokingdom; NA=North America Palaeokingdom; Cn=North China Palaeoarea; Cs=South China Palaeoarea; Gn=Nothoafroamerica Palaeoarea; Ga=Australoindia Palaeoarea; Gp=Palaeoantarctica Palaeoarea.

4.5.1 Distribution

The Cathaysia Palaeokingdom is best represented along the northern shores of the Tethys, in present-day China, Japan, Korea, Indo-China, Malaysia and Indonesia (Li & Yao 1985). In the Upper Permian, however, it is also found further west in the present-day Middle East and Turkey (Archangelsky & Wagner 1983, El-Khayal & Wagner 1985). The Turkish assemblages are of particular interest in that they include a mixture of Cathaysian and Gondwanan taxa, possibly reflecting an ecotone

Palaeokingdom	Gondwana			North America	Eurameria
Palaeoarea	Palaeoantarctica	Australoindia	Nothoafroamerica		
Sphenophyllum		1 2	1 2	1	1
Pterophyllum		2		1	1
Annularia			1	1	1
Cyathocarpus			1 2		1
Pseudoctenis		1 2			
Glossopteris	1 2	1 2	1 2		
Gangamopteris	1	1	1		
Buriadia		1	1		
Botrychiopsis		1	1		
Phyllotheca		1 2	1		
Rhipidopsis		1 2	1		
Palaeovittaria		2	1		
Cyclodendron		2	2		
Stellotheca		1			
Walkomiella		1			
Rhabdotaenia		1 2			
Schizoneura		1 2			
Raniganjia		1 2			
Pteronilssonia		2			

1 - Lower Permian; 2 - Upper Permian

Palaeokingdom	Gondwana			North America	Eurameria
Palaeoarea	Palaeoantarctica	Australoindia	Nothoafroamerica		
Nothorhacopteris			1		
Ferugliocladus			1		
Ugartecladus			1		
Brasilodendron			1		
Dizeugotheca			1 2		
Tinsleya				1	
Zeilleropteris				1	
Evolsonia				1	
Delnortia				1	
Rusellites				1	
Glenopteris				1	
Supaia				1	
Discinites				1	
Weissites					1
Dicksonites					1
Neurocallipteris					1
Rhaciphyllum					1
Culmitzschia					1 2
Dicranophyllum					1

1 - Lower Permian; 2 - Upper Permian

Fig. 4.7 — Occurrence of key form-genera in Permian palaeoareas. See text for main sources of data.

Palaeokingdom	North America	Eurameria	Cathaysia		Angara	
Palaeoarea			N. China	S. China	Kazakhstania	Pechora
Cathaysiopteris	1		1			
Autunia	1	1	1			
Taeniopteris	1	1 2	1 2	1 2		
Gigantoclea	1		1 2	1 2		
Gigantopteridium	1			1		
Walchia	1	1	2	2	1	1 2
Lobatannularia	1		1 2	2		
Cordaites	1	1			1 2	1 2
Odontopteris		1	1			
Calamites		1	1			
Asterophyllites		1	1 2			
Cyathocarpus		1	1 2	1 2		
Plagiozamites		2		1 2		
Ullmannia		2	2			
Sphenobaiera		2	2			
Quadrocaldus		2			2	
Neocalamites		2				
Lepidopteris		2				
Pseudovoltzia		2				
Schizoneura			2			

1 - Lower Permian; 2 - Upper Permian

Palaeokingdom	Cathaysia		Angara			
Palaeoarea	N. China	S. China	Kazakhstania	Pechora	Kuznetsk	Far East
Sphenophyllum	1 2	1 2	1 2	1 2	1	
Rhipidopsis	2	2		2	2	2
Annularia	1 2	2	2		1 2	2
Taeniopteris	1 2	1 2				2
Lobatannularia	1 2	2				2
Alethopteris	1					2
Tingia	1 2	1 2				
Oligocarpia	1 2			2		
Emplectopteris	1	1				
Lepidodendron	1	1 2				
Emplectopteridium	1					
Neuropteris	1					
Psygmophyllum	2	2	1 2	2		
Neuropteridium	2	2				2
Gigantopteris	2	1 2				
Fascipteris	2	2				
Dictyocallipteridium		1				
Rajahia		1 2				
Compsopteris		2		2	2	2
Otofolium		2				

1 - Lower Permian; 2 - Upper Permian

Fig. 4.7 — Occurrence of key form-genera in Permian palaeoareas. See text for main sources of data.

Palaeokingdom	Cathaysia		Angara			
Palaeoarea	N. China	S. China	Kazakhstania	Pechora	Kuznetsk	Far East
Pterophyllum	1					
Phyllotheca			1	1 2	1 2	2
Cordaites			1 2	1 2	1 2	2
Rufloria			1	1 2	1 2	2
Paracalamites			1 2	1 2	1 2	2
Zamiopteris			1	2	1 2	2
Pursongia			2		1 2	2
Nephropsis			1	1 2	1 2	
Annulina			1	1 2	1	
Dicranophyllum			1		1	
Mauerites			1			
Sylvopteris			1			
Biarmopteris			1			
Entsovia			1 2			
Viatcheslavia			2	1 2		
Fefilopteris			2		2	
Phylladoderma			2	2		
Rhaphidopteris			2	2		
Thamnopteris			2			
Mostotchkia			2			

1 - Lower Permian; 2 - Upper Permian

Palaeokingdom	Cathaysia		Angara			
Palaeoarea	N. China	S. China	Kazakhstania	Pechora	Kuznetsk	Far East
Tatarina			2			
Permotheca			2			
Tschernovia				1 2	1	
Prynadaeopteris				1 2	1 2	
Orthotheca				1 2		
Comia				2	2	2
Wattia				2		
Syniopteris				2		
Tundrodendron				2		
Lophoderma					1	
Angaropteridium					1	
Evenkiella					1	
Barakaria					1	
Glottophyllum					1 2	2
Angarodendron					1	
Paraschizoneura					2	
Listrophyllum					2	
Tychtopteris					2	2
Iniopteris					2	2
Yavorskyia					2	2

1 - Lower Permian; 2 - Upper Permian

Fig. 4.7 — Occurrence of key form-genera in Permian palaeoareas. See text for main sources of data.

between the two palaeokingdoms. There are also records of mixed assemblages in Uzbekistan and southernmost Kazakhstan, this time involving Angaran elements (see section 4.3.3.2).

4.5.2 Floristic history

Climatic conditions in Cathaysia appear to have continued virtually unaltered from the Carboniferous to the Permian. The Early Permian floras thus resembled those of the Upper Carboniferous more than those of the Permian of Laurasia. Conifers, cordaites and peltasperm pteridosperms made little significant headway into the lowland areas, but lycophytes, sphenophytes and pteridosperms remained dominant. A number of distinctively Cathaysian elements also occur here, such as the Gigantonomiales ('gigantopterids' *auct.*), noeggerathialean-like progymnosperms and plants with cycad-like foliage. Towards the end of the Permian, there is some evidence of plants similar to those found in Angara (Meyen 1987), although details of these remain unclear. Coal-forming peats continued to develop right up to the end of the Permian.

4.5.3 Palaeoareas

Subsidiary phytochoria have been recognized in the Cathaysia Palaeokingdom by Li & Yao (1985), referred to here as North China and South China palaeoareas, with a geographical boundary along the Kunlun–Qinling Mountains. It has been widely argued that this reflects a separation of North China and South China palaeocontinents (Wang 1985), although the evidence now suggests that the accretion of these 'blocks' had already occurred in the Devonian (Laveine *et al.* 1987). It may alternatively reflect the differing environmental settings suggested by the sedimentology: the North China sequences are predominantly non-marine, while the South China sequences include a high proportion of marine and pyroclastic deposits. There are also significant palaeolatitudinal differences between the palaeoareas.

4.5.3.1 North China Palaeoarea

Taphocoenoses of this palaeoarea occur extensively through northern China, and include the well-known Shanxi 'floras' (Lee 1964). Lower Permian assemblages are characterized by *Emplectopteris* and *Cathaysiopteris*, whilst the Upper Permian has yielded abundant *Taeniopteris*. *Gigantopteris* is restricted to rare occurrences in the Upper Permian.

4.5.3.2 South China Palaeoarea

South of the present-day Kunlun–Qinling Mountains, e.g. in Hunan and Yunnan, *Gigantopteris* occurs abundantly. However, *Emplectopteris* and *Cathaysiopteris* are rare in the Lower Permian, and *Taeniopteris* is rare in the Upper Permian. The Upper Permian of the South China Palaeoarea also has a number of endemic form-genera, including *Rajahia* and *Otofolium*. Li & Yao (1985) note differences between the palaeoareas at the species level, but this is beyond the scope of the present discussion. Meyen's (1987) claim that *Gigantonoclea* is restricted to this palaeoarea is contradicted by Li & Yao (1985).

4.6 NORTH AMERICA PALAEOKINGDOM

This phytochorion was once thought to have Cathaysian affinities (White 1912, Darrah 1937) but subsequent work (e.g. Mamay 1968) has shown that the similarities are superficial. It may have evolved out of the Cordillera Palaeoarea, recognized by Read (in Read & Mamay 1964) for so-called 'upland' Upper Carboniferous assemblages in the Eurameria Palaeokingdom, although their exact relation remains unclear.

4.6.1 Distribution

So far, this palaeokingdom is best represented in the Lower Permian of present-day western North America, representing the western margins of Laurasia. Plant fossils reported (but as yet not fully described) from the Permian of Venezuela might also belong here (Odreman & Wagner in press).

4.6.2 Floristic history

This phytochorion is believed to represent floras growing in semi-arid environments (Meyen 1987). It is still relatively poorly known but, like the other Permian palaeoequatorial phytochoria, it appears to include a high proportion of conifers and ?marattialean ferns. The presence of apparently endemic ?progymnosperms, peltasperm pteridosperms and cycad-like plants, however, gives these assemblages a most distinctive character (Read & Mamay 1964, Mamay 1968, 1976). Also present is foliage very similar to the gigantopterids of the Cathaysia Palaeokingdom. Gao & Thomas (1989) have argued that, despite the wide geographical separation between Cathaysia and North America in the Permian, the transportation of seeds by oceanic currents could easily explain the apparent similarity in the vegetation. Wagner in El-Khayal & Wagner (1985) has also implied that it reflects palaeoclimatic similarities between the two areas, but this appears to be incompatible with the reputed semi-arid environment of the North America Palaeokingdom (see above). An alternative scenario is that these fossils represent groups of plants that were widely distributed in the extra-basinal habitats in the Carboniferous, and that only in the Permian did they selectively invade the sediment-accumulating areas. Unfortunately, very little is known of the fructifications of the plants that bore these leaves, in either Cathaysia or North America. It is thus impossible to be certain whether the similarity of the leaves reflects homology in distantly related groups or true phylogenetic affinity. It cannot be denied, however, that the morphological similarity is most striking.

4.6.3 Palaeoareas

Read & Mamay (1964) suggest that distinct *Supaia*, *Glenopteris* and '*Gigantopteris*' 'floras' could be recognized within the North America Palaeokingdom. However, the stratigraphic control on many of the localities is poor and it is possible that at least some of the apparent variation is due to differences in stratigraphical position. These phytochoria are thus not dealt with further here.

4.7 CONCLUDING REMARKS

It is evident from Fig. 4.1 that there is a significant increase in palaeophytogeographic provincialism towards the top of the Palaeozoic. The Angara Palaeokingdom is already established by the base of the Carboniferous, the Gondwana Palaeokingdom becomes distinct in the Upper Carboniferous, and the Cathaysia and North America palaeokingdoms in the Permian. The progressive appearance of these palaeokingdoms is indicated by the presence of a Gondwana Palaeoarea in the Lower Carboniferous and Cathaysia and Cordillera palaeoareas in the Upper Carboniferous. This model is contrary to that of Chaloner & Creber (1988), who suggest, on the basis of distance measurements of Permian assemblage composition, that the Gondwana Palaeokingdom became established before the Angara Palaeokingdom. It has to be recognized, however, that the evidence from the Lower Carboniferous of Gondwana is poor, and might not be giving a true reflection of when this phytochorion became established as a full palaeokingdom.

As already hinted, the increase in provincialism was probably a consequence of the development of polar ice-caps (John 1979), which in turn produced a greater palaeoclimatic zonation over the world. By the lower Mesozoic, however, this palaeophytogeographic provinciality had rapidly broken down (Chapter 6) as a response to palaeoclimatic changes. Not until the Upper Jurassic does a comparable palaeophytogeography become re-established. This explains why in no other part of the geological column (at least below the Tertiary) have plant fossils played such a significant role in the establishment of palaeocontinental reconstructions.

REFERENCES

Abdulin, A. A., Bikova, M. C., Kushev, G. L., Bublichenko, N. L., Nasikanova, O. N. & Radchenko, M. I. (eds) (1975) The Carboniferous of Kazakhstan. *Satpaev Institute of Geological Sciences, Alma-Ata* [In Russian, with English summary].

Allen, K. C. & Dineley, D. L. (1988) Mid-Devonian to mid-Permian floral and faunal regions and provinces. In: Harris, A. L. & Fettes, D. J. (eds) *The Caledonian–Appalachian Orogeny. Geol. Soc. London, Spec. Publ.* **38** 531–548.

Archangelsky, S. (1990) Plant distribution in Gondwana during the Late Palaeozoic. In: Taylor, T. N. & Taylor, E. L. (eds) *Antarctic paleobiology*. Springer, Berlin, pp. 102–117.

Archangelsky, S. & Arrondo, O. G. (1969) The Permian taphofloras of Argentina with some considerations about the presence of 'northern' elements and their possible significance. In: *Gondwana stratigraphy. IUGS Symposium, Buenos Aires (Mar del Plata), October 1967.* UNESCO, pp. 197–212.

Archangelsky, S. & Arrondo, O. G. (1975) Paleogeografá y plantas fósiles en el Pérmico inferior Austrosudamericano. *Actas I Congr. Arg. de Pal. y Bioestrat.* (Tucumán, 1974) **1** 479–496.

Archangelsky, S. & Wagner, R. H. (1983) *Glossopteris anatolica* sp. nov. from uppermost Permian strata in south-east Turkey. *Bull. Br. Mus. nat. Hist. (Geol.)* **37** 81–91.

Arnold, C. A. (1941) Some Paleozoic plants from central Colorado and their stratigraphic significance. *Contr. Mus. Paleontol. Univ. Michigan* **6** 59–70.

Asama, K. (1984) *Gigantopteris* flora in China and southeast Asia. In: Kobayashi, T., Toriyama, R. & Hashimoto, W. (eds) *Geology and palaeontology of southeast Asia*. pp. 311–323.

Chaloner, W. G. & Creber, G. T. (1988) Fossil plants as indicators of late Palaeozoic plate positions. In: Audley-Charles, M. G. & Hallam, A. (eds) *Gondwana and Tethys. Geol. Soc. Lond., Spec. Publ.* **37** 201–210.

Chaloner, W. G. & Lacey, W. S. (1973) The distribution of Late Palaeozoic floras. In: Hughes, N. F. (ed.) *Organisms and continents through time. Spec. Pap. Palaeont.* **12** 271–289.

Chaloner, W. G. & Meyen, S. V. (1973) Carboniferous and Permian floras of the northern continents. In: Hallam, A. (ed.) *Atlas of palaeobiogeography*. Elsevier, Amsterdam, pp. 169–186.

Darrah, W. C. (1937) Some floral relationships between the late Paleozoic of Asia and North America. *Problems in palaeontology, Moscow Univ., Lab. Palaeont. Publ.* **2–3** 195–205.

Doubinger, J. (1956) Contribution à l'étude des flores autuno-stépheniennes. *Mém. Soc. Géol. France* (N.S.) **35** 1–180.

Durante, M. V. (1976) Paleobotanicheskoe obocnovanie stratigrafii Karbona i Permi Mongolii. *Trudy Sovmestn. Sov.-Mongol. Exped.* **19** 5–279 [In Russian].

Durante, M. V. (1983) Existence of an Upper Permian mixed Cathaysio-Angaran flora in Nanshan (north China). *Geobios* **16** 241–242.

El-Khayal, A. A. & Wagner, R. H. (1985) Upper Permian stratigraphy and megafloras of Saudi Arabia: palaeogeographic and climatic implications. *C. R. 10e Congr. Int. Strat. Géol. Carbon.* (Madrid, 1983) **3** 17–26.

Fefilova, L. A. (1978) *Listostebel'nye mkhi Permi Evropeiskogo Severa SSSR.* Komi Branch of the Geological Institute, Leningrad [In Russian].

Gao Z. & Thomas, B. A. (1989) A review of fossil cycad megasporophylls, with new evidence of *Crossozamia* Pomel and its associated leaves from the Lower Permian of Taiyuan, China. *Rev. Palaeobot. Palynol.* **60** 205–223.

Gastaldo, R. (1987) Confirmation of Carboniferous clastic swamp communities. *Nature* **326** 869–871.

Gee, C. T. (1989) Permian *Glossopteris* and *Elatocladus* megafossil floras from the English Coast, eastern Ellsworth Land, Antarctica. *Antarctic Sci.* **1** 35–44.

Gomankov, A. V. & Meyen, S. V. (1986) Tatarinovaya flora (sostav i rasprostpanenie v pozdnei Permi Evrazii). *Trudy Geol. Inst. Akad. Nauk SSSR* **401** 1–175.

Gorelova, S. G., Men'shikova, L. V. & Khalfin, L. L. (1973) Fitostratigrafiya i opredelitel' pactenii verkhnepaleozoiskikh uglenosnykh otlozhenii Kuznetskogo Basseina. *Trudy S.N.I.I.G.G.I.M.Ca.* **140** 1–169 [In Russian].

Gothan, W. (1951) Die merkwürdigen pflanzengeographischen Besonderheiten in den mitteleuropäischen Karbonfloren. *Palaeontographica B* **91** 109–130.

Gu & Zhi (1974) *Palaeozoic plants from China.* Scientific Press, Beijing [In Chinese].

Hallam, A. (1983) Supposed Permo-Triassic megashear between Laurussia and Gondwana. *Nature* **301** 499–502.

Harland, W. B., Cox, A. V., Llewellyn, P. G., Pickton, C. A. G., Smith, A. G. & Walters, R. (1982) *A geologic time scale.* University Press, Cambridge.

John, B. (1979) The Great Ice: Permo-Carboniferous. In: John, B. (ed.) *The winters of the world. Earth under the ice ages.* David & Charles, Newton Abbott, pp. 154–172.

Jongmans, W. J. (1939) Die Kohlenbecken des Karbons und Perms im USSR und Ost-Asien. *Geol. Sticht.* **1934–1937** 15–192.

Jongmans, W. J. (1952) Some problems on Carboniferous stratigraphy. *C. R. 3e Congr. Int. Strat. Géol. Carbon.* (Heerlen, 1951) **1** 295–306.

Kerp, J. H. F. & Fichter, J. (1985) Die Makrofloren des saarpfälzischen Rotliegenden (? Ober-Karbon — Unter Perm; SW-Deutschland). *Mainzer geowiss. Mitt.* **14** 159–286.

Laveine, J.-P., Lemoigne, Y., Li X., Wu X., Zhang S., Zhao X., Zhu W. & Zhu J. (1987) Paléogéographie de la Chine au Carbonifère à la lumière des données paléobotaniques, par comparaison avec les assemblages carbonifères d'Europe occidentale. *C. R. Acad. Sci. Paris* (Sér. II) **304** 391–394.

Leary, R. L. (1974) Reconstruction of the coal age uplands. *The Explorer* **16** 27–29.

Leary, R. L. (1981) Early Pennsylvanian geology and paleobotany of the Rock Island County, Illinois Area. Part 1: Geology. *Ill. State Mus. Repts Investigat.* **37** 1–88.

Lee, H. H. (1964) The succession of Upper Palaeozoic plant assemblages of northern China. *C. R. 5e Congr. Int. Strat. Géol. Carbon.* (Paris, 1963) **2** 531–537.

Leeder, M. R. (1988) Recent developments in Carboniferous geology: a critical review with implications for the British Isles and N. W. Europe. *Proc. Geol. Ass.* **99** 73–100.

Li X. & Yao Z. (1985) Carboniferous and Permian floral provinces in East Asia. *C. R. 9e Congr. Int. Strat. Géol. Carbon.* (Washington & Urbana 1979) **5** 95–101.

Lyons, P. C. & Darrah, W. C. (1989) Earliest conifers of North America: upland and/or paleoclimatic indicators? *Palaios* **4** 480–486.

Mamay, S. H. (1968) *Russellites*, new genus, a problematical plant from the Lower Permian of Texas. *U.S. Geol. Surv. Prof. Pap.* **593** I1–I15.

Mamay, S. H. (1976) Paleozoic origin of cycads. *U.S. Geol. Surv. Prof. Pap.* **934** 1–48.

Meyen, S. V. (1969) New data on relationship between Angara and Gondwana Late Palaeozoic floras. In: *Gondwana stratigraphy, IUGS Symposium, Buenos*

Aires, 1967. UNESCO, Paris, pp. 141–157.

Meyen, S. V. (1982) The Carboniferous and Permian floras of Angaraland. (A synthesis). *Biol. Mem.* **7**, 1–109.

Meyen, S. V. (ed.) (1983) *Paleontologicheskii atlas Permskikh otlozhenii Pechorskogo ugol'nogo bacceina*. Komi Branch of the Geological Institute, Leningrad [In Russian].

Meyen, S.V. (1987) *Fundamentals of palaeobotany*. Chapman & Hall, London.

Neuburg, M. F. (1948) *Verkhnepaleozoiskaya flora Kuznetskogo basseina*. Institute of Palaeontology, USSR Academy of Sciences, Moscow (*The palaeontology of the USSR*, vol. 12, pt 3(2)) [In Russian].

Neuburg, M. F. (1961) Present state of the question on the origin, stratigraphic significance and age of Paleozoic floras of Angaraland. *C. R. 4e Congr. Int. Strat. Géol. Carbon*. (Heerlen, 1958) **2** 443–452.

Niklas, K. J., Tiffney, B. H. & Knoll, A. H. (1980) Apparent changes in the diversity of fossil plants. A preliminary assessment. In: Hecht, M. K., Steere, W. C. & Wallace, B. (eds) *Evolutionary biology*, vol. 12. Plenum Press, New York, pp. 1–89.

Odreman, O. & Wagner, R. H. (in press) Notes on Carboniferous and Permian floras in the Venezuelan Andes. *Proceedings of the IUGS–SCCS Field and General Meeting* (Turkey, 1978).

Oshurkova, M.V. (1967) *Paleofitologicheskoe obosnovanie stratigrafii verkhnikh svit Kamennoygol'nikh otlozhenii Karagandinskogo Basseina*. All Union Geological Research-Science Institute, Leningrad [In Russian].

Pfefferkorn, H. W. & Gillespie, W. H. (1980) Biostratigraphy and biogeography of plant compression fossils in the Pennsylvanian of North America. In: Dilcher, D. L. & Taylor, T. N. (eds) *Biostratigraphy of fossil plants*. Dowden, Hutchinson & Ross, Stroudsburg, Pennsylvania, pp. 93–118.

Pfefferkorn, H. W. & Gillespie, W. H. (1982) Plant megafossils near the Mississippian–Pennsylvanian boundary in the Pennsylvanian System stratotype, West Virginia/Virgina. In: Ramsbottom, W. H. C., Saunders, W. B. & Owens, B. (eds) *Biostratigraphic data for a Mid- Carboniferous boundary*. Subcommission on Carboniferous Stratigraphy, Leeds, pp. 128–133.

Phillips, T. L. (1980) Stratigraphic and geographic occurrences of permineralized coal-swamp plants — Upper Carboniferous of North America and Europe. In: Dilcher, D. L. & Taylor, T. N. (eds) *Biostratigraphy of fossil plants*. Dowden, Hutchinson & Ross, Stroudsburg, Pennsylvania, pp. 25-92.

Phillips, T. L. & Peppers, R. A. (1984) Changing patterns of Pennsylvanian coal-swamp vegetation and implications of climatic control on coal occurrence. *Int. J. Coal Geol*. **3**, 205–-55.

Plumstead, E.P. (1967) A review of contributions to the knowledge of Gondwana mega-plants and floras of Africa published since 1950. In: *IUGS Reviews, prepared for the First Symposium on Gondwana Stratigraphy, 1967, Haarlem, Netherlands*. IUGS Secretariat, pp. 139–148.

Plumstead, E. P. (1973) The late Palaeozoic *Glossopteris* flora. In: Hallam, A. (ed.) *Atlas of palaeobiogeography*. Elsevier, Amsterdam, pp. 187–205.

Radchenko, M. I. (1967) *Kamennougol'naya flora yugo-vostochnogo Kazakhstana*. Satpaev Institute of Geological Sciences, Alma-Ata [In Russian].

Radchenko, M. I. (1985) *Atlas (opredelitel') Kamennougol'noi flory Kazakhstana*. Satpaev Institute of Geological Sciences, Alma-Ata [In Russian].

Raymond, A. (1985) Floral diversity, phytogeography, and climatic amelioration during the Early Carboniferous (Dinantian). *Paleobiology* **11** 293–309.

Raymond, A., Parker, W. C. & Parrish, J. T. (1985) Phytogeography and paleoclimate of the Early Carboniferous. In: Tiffney, B. H. (ed.) *Geological factors and the evolution of plants*. University Press, Yale, pp. 169–222.

Read, C. B. (1955) Floras of the Pocono Formation and Price Sandstone in parts of Pennsylvania, Maryland, West Virginia and Virginia. *U.S. Geol. Surv. Prof. Pap.* **263**, 1–32.

Read, C. B. & Mamay, S. H. (1964) Upper Paleozoic floral zones and floral provinces of the United States. *U.S. Geol. Surv. Prof. Pap.* **454** K1–K35.

Remy, R. & Remy, W. (1959) Beiträge zur Kenntnis der Rotliegend-Flora Thüringens. IV. *Dtsch. Akad. Wiss. Berlin, Kl. Chem. geol. und Biol.* **2**.

Retallack, G. J. (1980) Late Carboniferous to Middle Triassic megafossil floras from the Sydney Basin. In: Herbert, C. & Helby, R. J. (eds) *A güide to the Sydney Basin. Geol. Surv. NSW Bull.* **26** 384–430.

Rigby, J. F. & Schopf, J. M. (1969) Stratigraphic implications of Antarctic paleobotanical studies. In: *Gondwana stratigraphy*. (IUGS Symposium, Buenos Aires, 1967). UNESCO, Paris, pp. 91–106.

Rowley, D. B., Raymond, A., Parrish, J. T., Lottes, A. L., Scotese, C. R. & Ziegler, A. M. (1985) Carboniferous paleogeographic, phytogeographic, and paleoclimatic reconstructions. *Int. J. Coal Geol.* **5** 7–42.

Schopf, J. M. & Askin, R. A. (1980) Permian and Triassic floral biostratigraphic zones of southern land masses. In: Dilcher, D. L. & Taylor, T. N. (eds) *Biostratigraphy of fossil plants*. Dowden, Hutchinson & Ross, Stroudsburg, Pennsylvania, pp. 119–152.

Schweitzer, H.-J. (1986) The land flora of the English and German Zechstein sequences. In: Harwood, G. M. & Smith, D. B. (eds) *The English Zechstein and related topics. Geol. Soc. Lond., Spec. Publ.* **22** 31–54.

Scotese, C. R. & McKerrow, W. S. (1990) Revised world maps and introduction. In: McKerrow, W. S. & Scotese, C. R. (eds). *Palaeozoic palaeogeography and biogeography. Geol. Soc. Lond. Mem.* **12** 1–21.

Scott, A. C. (1985) Distribution of Lower Carboniferous floras in northern Britain. *C. R. 9e Congr. Int. Strat. Géol. Carbon.* (Washington & Urbana, 1979) **5** 77–82.

Scott, A. C. & Chaloner, W. G. (1983) The earliest fossil conifer from the Westphalian B of Yorkshire. *Proc. Roy. Soc. Lond. B* **220** 163–182.

Sixtel, T.A. (ed.) (1975) *Biostratigrafiya verkhnego paleozoya gornogo obramleniya*

Yuznoi Fergany. FAN, Tashkent [In Russian].

Sixtel, T. A. & Savitskaya, L. I. (1985) Phytostratigraphic data and the palaeofloristic boundaries in Central Asia during the Carboniferous Period. *C. R. 10e Congr. Int. Strat. Gol. Carbon*. (Madrid, 1983) **2** 371–374.

Surange, K. R. (1975) Indian Lower Gondwana floras: a review. In: Campbell, K. S. W. (ed.) *Gondwana geology*. Australian National University Press, pp. 135–147.

Tarling, D. H. (1980) Upper Palaeozoic continental distributions based on palaeomagnetic studies. In: Panchen, A. L. (ed.) *The terrestrial environment and the origin of land vertebrates. Systematics Association, Spec. Vol. 15* 11–37.

Tarling, D. H. (1985) Carboniferous reconstructions based on palaeomagnetism. *C. R. 10e Congr. Int. Strat. Géol. Carbon*. (Madrid, 1983) **3** 153–162.

Tidwell, W. D. (1967) Flora of the Manning Canyon Shale, Part 1: A lowermost Pennsylvanian flora from the Manning Canyon Shale, Utah, and its stratigraphic significance. *Brigham Young Univ. Geol. Stud.* **14** 1–63.

Vakhrameev, V. A., Dobruskina, I. A., Meyen, S. V. & Zaklinskaja, E. D. (1978) *Paläozoische und mesozoische Floren Eurasiens und die Phytogeographie dieser Zeit*. Gustav Fischer, Jena (Russian edition published in 1970 by the Geological Institute USSR, Moscow).

Wagner, R. H. (1982) Floral changes near the Mississippian–Pennsylvanian boundary; an appraisal. In: Ramsbottom, W. H. C., Saunders, W. B. & Owens, B. (eds) *Biostratigraphic data for a Mid-Carboniferous boundary*. Subcommission on Carboniferous Stratigraphy, Leeds, pp. 120–127.

Wagner, R. H. (1985) Megafloral zones of the Carboniferous. *C. R. 9e Congr. Int. Strat. Géol. Carbon*. (Washington & Urbana, 1979) **2** 109–134.

Wagner, R. H., Soper, N. J. & Higgins, A. K. (1982) A Late Permian flora of Pechora affinity in north Greenland. *Grønlands Geol. Unders.* **108** 5–13.

Wagner, R. H., Winkler Prins, C. F. & Granados, L. F. (eds) (1983) *The Carboniferous of the world. I. China, Korea, Japan & S.E. Asia.* Instituto Geológico y Minero de España, Madrid.

Wagner, R. H., Winkler Prins, C. F. & Granados, L. F. (eds) (1985) *The Carboniferous of the world. II. Australia, Indian Subcontinent, South Africa, South America & North Africa.* Instituto Geológico y Minero de España, Madrid.

Wang Z. Q. (1985) Palaeovegetation and plate tectonics: palaeophytogeography of North China during Permian and Triassic times. *Palaeogeogr., Palaeoclimatol., Palaeoecol.* **49** 25–45.

White, D. (1912) The characters of the fossil plant *Gigantopteris* Schenk and its occurrence in North America. *U.S. Natl Mus. Proc.* **41** 493–516.

Zhang, Z. M., Liou, J. C. & Coleman, R. G. (1984) An outline of the plate tectonics of China. *Bull. geol. Soc. Amer.* **95** 295–312.

Zhao X. & Wu X. (1985) Carboniferous macrofloras of south China. *C. R. 9e Congr. Int. Strat. Géol. Carbon.* (Washington & Urbana 1979) **5** 109–114.

Ziegler, A. M., Scotese, C. R., McKerrow, W. S., Johnson, M. E & Bambach, R. K. (1979) Paleozoic paleogeography. *Ann. Rev. Earth Planet. Sci.* **7** 473–502.

5

Carboniferous and Permian biostratigraphy

C. J. Cleal

There has been more work on plant biostratigraphy in the Carboniferous and Permian than in any other part of the geological column. Particularly in the non-marine facies of the Eurameria Palaeokingdom, they provide correlations that rival any produced by animal fossils or plant microfossils. This chapter will review the various biostratigraphic schemes that are available in each of the palaeokingdoms, and how they compare with competing schemes for non-marine sequences established using other groups of organisms.

5.1 EURAMERIA PALAEOKINGDOM

Carboniferous and Permian plant biostratigraphy in Europe dates back to the late nineteenth century, with the earliest formal biozonation being established by Zeiller (1894) for the Westphalian Series. This work has been further developed by Bertrand (1914, 1919), Dix (1934), Corsin & Corsin (1971), Paproth *et al.* (1983) and Laveine (1986). In North America, biozonal schemes were proposed by Bell (1938) and, perhaps more significantly, by Read & Mamay (1964). In that part of the USSR within the Eurameria Palaeokingdom, schemes have been proposed by Novik (1952, 1978).

The most significant advance in recent years has been the classification proposed by Wagner (1984 — hereafter referred to as the Wagner Classification). It covers the entire Carboniferous and, although it was established largely on data from western and central Europe, was intended for use throughout the palaeoequatorial belt. As pointed out by Fissunenko & Laveine (1984), there are discrepancies in the stratigraphic ranges of certain plant species between western Europe and the Donets and Caucasus basins, and this could cause difficulties in applying the Wagner Classification to these more eastern areas. Nevertheless, Wagner was able to establish a broad correlation between at least some of his biozones and those established by Novik (1952) for the Donets sequence.

5.1.1 Europe Palaeoarea

Within this palaeoarea, fossils representing three distinct habitats have been recognized: adpressions found in mainly grey, Carboniferous sediments, probably representing riparian or lake side vegetation; coal-ball petrifactions, representing the main swamp-forests of the Carboniferous; and adpressions from Upper Carboniferous and Permian red-beds, representing so-called 'upland' or extra-basinal vegetation. The petrifactions from the Lower Carboniferous (reviewed by Scott *et al.* 1984) probably represent a fourth habitat-type but, as they have not been subject to any detailed biostratigraphic analysis, they will not be considered further here.

5.1.1.1 Carboniferous adpressions

The Wagner Classification is the most refined biostratigraphy available for these fossils, the Carboniferous being divided into 16 assemblage biozones (Fig. 5.1). Space does not permit the reproduction here of the comprehensive set of range charts provided by Wagner, but Figs 5.2–5.7 summarize the ranges of the key taxa for recognizing the biostratigraphic boundaries. It should be emphasized, however, that Wagner's charts remain the most useful source of biostratigraphic data on these fossils, and should be used in conjunction with the information provided here. The biostratigraphy summarized in this chapter has been modified slightly from Wagner's original scheme, as follows.

(a) Wagner correctly pointed out that the changes in the plant fossil assemblages are gradational and that the choice of biozone boundaries is therefore somewhat arbitrary. However, there are certain levels at which a significant number of species appear and/or disappear, and which provide useful biostratigraphic markers. Some of these are not recognized in Wagner's original scheme, but could constructively be introduced to improve its resolution. In order not to disrupt Wagner's original classification, however, these have been largely recognized as subbiozones.
(b) Cleal (1984) proposed a minor change to the biostratigraphy of the Westphalian D, by recognizing a tripartite division, rather than Wagner's original bipartite split. In order, again, to maintain the basic integrity of Wagner's original classification, it is proposed to revert to just two biozones, but to split the upper one into subbiozones.
(c) The boundary between the *L. rugosa* and *P. linguaefolia* biozones, as originally defined by Wagner, is difficult to recognize on purely palaeobotanical grounds. It is therefore proposed to place it at a slightly lower level, where there is a more marked palaeobotanical change (see Figs 5.4–5.5).
(d) Several of Wagner's biozones were named using two species. In order to simplify the nomenclature, these have been renamed using just one of the species.

The boundaries between the biozones are marked by the appearance and/or extinction of one or more taxa, and are thus assemblage zones in the normally accepted sense (Hedberg 1976, Holland *et al.* 1978). Consequently, the assignment of a particular fossil assemblage in isolation may be difficult, although usually not

Series	Stages	Subbiozones	Biozones
NAMURIAN	Yeadonian	*Neuralethopteris larischii*	*Pecopteris aspera*
	Marsdenian	*Sigillaria elegans*	
	Kinderscoutian to Chokierian		*Lyginopteris larischii*
	Arnsbergian		*Lyginopteris stangeri*
	Pendleian	*Lyginopteris fragilis*	*Neuropteris antecedens*
VISEAN	Brigantian	*Diplopteridium*	
	Asbian - Chadian	*Spathulopteris*	*Triphyllopteris*
TOURNAIS -IAN	Ivorian	*Lepidodendropsis*	
	Hastarian		*Adiantites*

Fig. 5.1 — Plant biostratigraphy for the Eurameria Palaeokingdom, based mainly on the distribution of adpressions in the Europe Palaeoarea.

impossible provided it has a reasonable taxonomic diversity. The scheme really comes into its own, however, when dealing with a range of assemblages spread over a stratigraphic interval, such that the appearances and extinctions of the taxa can be observed. Examples of its application to such sequences can be found in Cleal (1984), Wagner *et al.* (1983) and Zodrow & Cleal (1985).

Series	Stages	Subbiozones	Biozones
LOWER PERMIAN	Autunian		*Autunia conferta*
STEPHANIAN	Stephanian C		*Sphenophyllum angustifolium*
	Stephanian B		*Alethopteris zeilleri*
	Baruellian		*Lobatopteris lamuriana*
	Cantabrian		*Odontopteris cantabrica*
WESTPHALIAN	Westphalian D	*Dicksonites plueckenetii*	*Lobatopteris vestita*
		Lobatopteris micromiltoni	
			Linopteris bunburii
	Bolsovian	*Alethopteris serlii*	*Paripteris linguaefolia*
		Laveineopteris rarinervis	
		Neuropteris semireticulata	
	Duckmantian	*Sphenophyllum majus*	*Lonchopteris rugosa*
		Neuropteris hollandica	
	Langsettian	*Laveineopteris loshii*	*Lyginopteris hoeninghausii*
		Neuralethopteris jongmansii	

Fig. 5.1 — Plant biostratigraphy for the Eurameria Palaeokingdom, based mainly on the distribution of adpressions in the Europe Palaeoarea.

The *Adiantites* Biozone represents assemblages transitional between those found in the Upper Devonian and the more typical Lower Carboniferous assemblages. Few examples of such transitional assemblages are known, the best documented being from the Pocono Formation in the Appalachian Mountains.

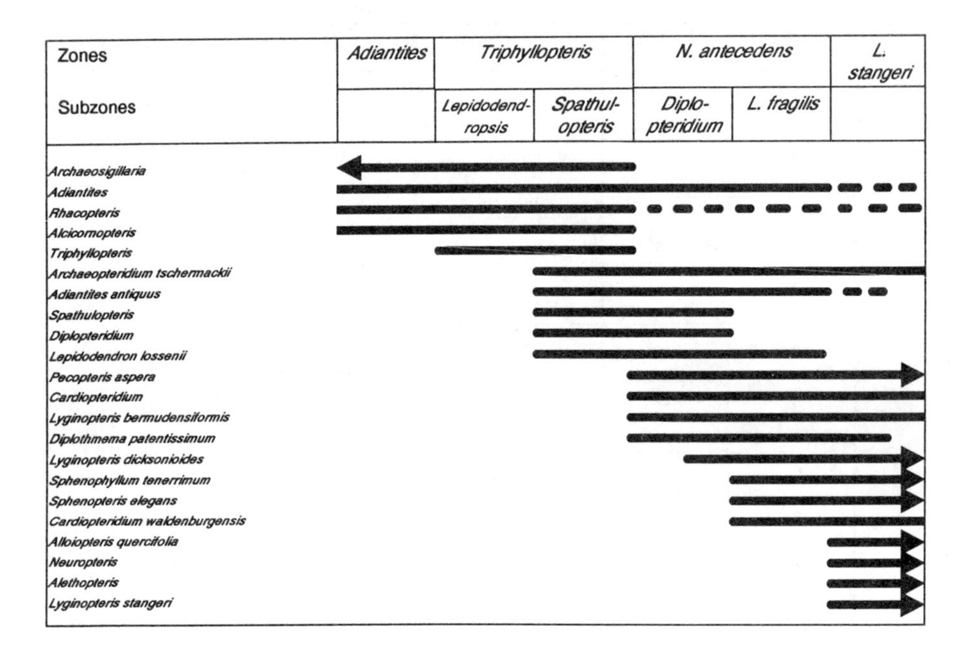

Fig, 5.2 — Stratigraphic distribution of key plant fossils in the Lower Carboniferous of the Europe Palaeoarea. See text for main sources of data.

The more typical Lower Carboniferous assemblages are represented by the *Triphyllopteris, N. antecedens* and *L. stangeri* biozones. They are known throughout the palaeoarea, and include some classic localities, such as the Burdiehouse Limestone in Scotland, Geigen near Hof in Bavaria, Rothwaltersdorf (now Czerwienczyce) in Lower Silesia, the Upper Silesia–Moravia Coalfield, and the Horton Group in Nova Scotia (Wagner 1984 provides a more complete review of the occurrences). Dominant elements tend to be the sphenophyte *Archaeocalamites*, ?progymnosperm foliage such as *Rhacopteris*, and a variety of early pteridosperm foliage such as *Sphenopteridium, Spathulopteris, Cardiopteridium* and *Lyginopteris*.

The *L. larischii* Biozone marks the end of the Lower Carboniferous. Many of the typically Visean taxa (e.g. *Sphenopteridium, Adiantites*) do not extend into this zone, whilst certain Upper Carboniferous forms make their first appearance (e.g. *Annularia, Alethopteris*). There are relatively few records of assemblages belonging to this biozone, the best being from the Upper Silesian–Moravia Coalfield. This makes it difficult to establish the exact details of the junction between this and the overlying biozone, which represents the 'floral-break' of Gothan (1952), purported to mark the boundary between the lower and upper Carboniferous 'floras'.

Most Namurian plant fossils belong to the *P. aspera* Biozone. They are very widely distributed throughout this palaeoarea, although assemblages tend to be of limited diversity. The biozone sees the gradual introduction of a series of taxa, generally regarded as typically Upper Carboniferous in aspect, such as *Paripteris* spp., *Sigillaria elegans, Asterophyllites grandis, Alethopteris lonchitica*.

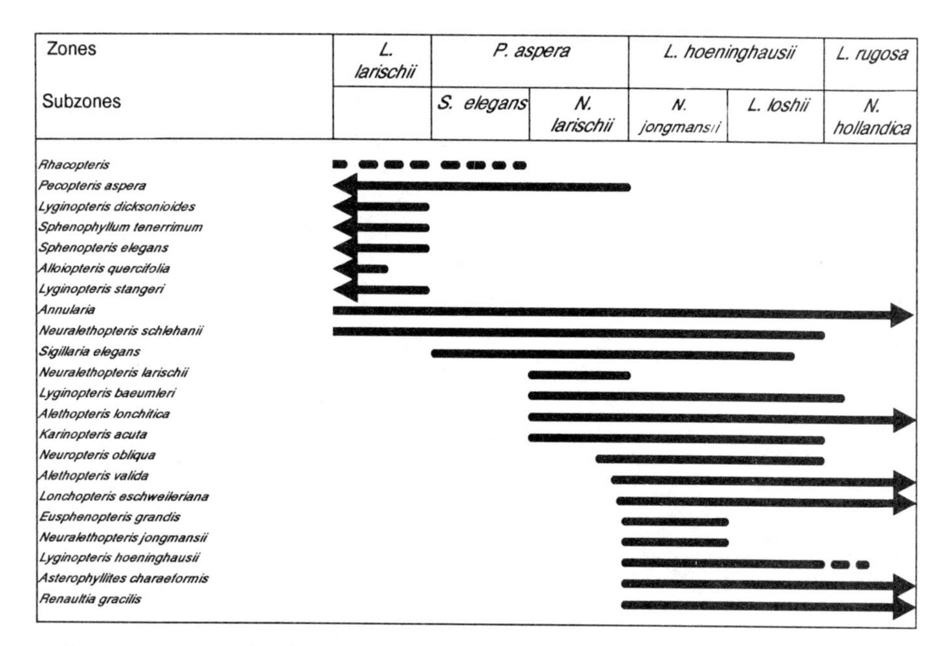

Fig.. 5.3 — Stratigraphic distribution of key plant fossils in the Namurian to mid-Westphalian of the Europe Palaeoarea. See text for main sources of data.

Zones
L. hoeninghausii
L. rugosa
P. linguaefolia
Subzones
N. jongmansii
L. loshii
N. hollandica
S. majus
N.semi-reticulata
L. rarinervis
Alethopteris valida
Lonchopteris eschweileriana
Asterophyllites charaeformis
Corynepteris angustissima
Eusphenopteris nummularia
Eusphenopteris trigonophylla
Eusphenopteris obtusiloba
Mariopteris muricata
Margaritopteris conwayi
Neuropteris heterophylla
Eusphenopteris sauveurii
Laveineopteris loshii
Alethopteris davreuxii
Lobatopteris miltoni
Eusphenopteris neuropteroides
Corynepteris coralloides
Zeilleria frenzlii
Eusphenopteris scribanii
Macroneuropteris scheuchzeri
Sphenophyllum myriophyllum
Renaultia rotundifolia
Renaultia chaerophylloides
Oligocarpia brongniartii

Fig. 5.4 — Stratigraphic distribution of key plant fossils in the lower to middle Westphalian of the Europe Palaeoarea. See text for main sources of data.

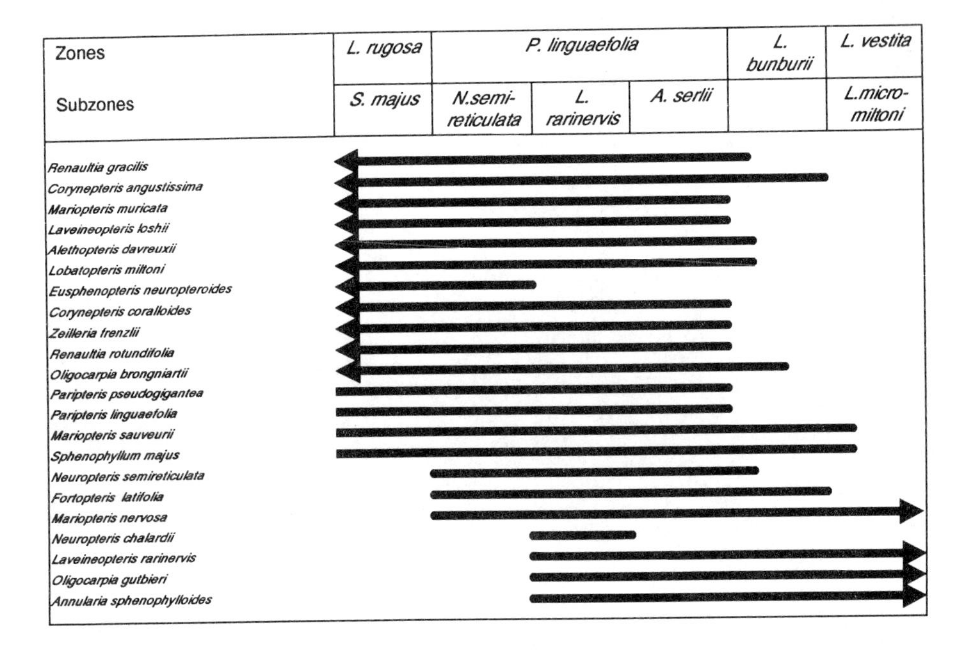

Fig. 5.5 — Stratigraphic distribution of key plant fossils in the middle to upper Westphalian of the Europe Palaeoarea. See text for main sources of data.

The *L. hoeninghausii, L. rugosa, P. linguaefolia* and *L. bunburii* biozones represent the classic 'Coal Measures floras' found throughout the palaeoarea. They include some of the most diverse plant fossil assemblages known from the Palaeozoic, and are generally dominated by lagenostomalean and trigonocarpalean pteridosperms, marattialean ferns, and calamostachyalean and bowmanitalean sphenophytes.

The *L. vestita* Biozone marks the transition between typically Westphalian and typically Stephanian assemblages. The effects of Variscan tectonic activity in the very late Westphalian, resulting from the collision between the Gondwana and Laurasia plates, had a major impact on the distribution of vegetation. In some areas, such as southern Britain, the Maritime Provinces of Canada, and the Zwickau and Saarland basins in Germany, the lowland swamp-forests continued to flourish, with the result that these areas yield well-developed *L. vestita* Biozone plant fossils. Elsewhere, uplift resulted in the migration of conifer-dominated vegetation (e.g. the Keele Formation of the English Midlands), whilst other areas became erosional and contain no strata of this age (e.g. Nord-Pas-de-Calais in northern France and the Ruhr in northern Germany).

Stephanian assemblages belong to the *O. cantabrica, L. lamuriana, A. zeilleri and S. angustifolium* biozones. The uplift and dissection of the landscape resulting from the Variscan tectonic processes means that these biozones are best represented in the small intra-montane basins in France and Germany. However, there are also records from northern Spain, the Appalachians and the Donets. They are mainly dominated by trigonocarpalean pteridosperms, marattialean ferns and bowmanitalean sphenophytes.

The base of the *A. conferta* Biozone was for many years regarded as marking the Carboniferous–Permian boundary. It is now clear, however, that some *A. conferta* Biozone assemblages are coeval with more typical upper Stephanian assemblages and that the junction reflects localized environmental changes (Broutin *et al.* 1990). Although retaining many of the characteristics of assemblages from the Stephanian coal-bearing deposits, it sees the introduction of peltaspermalean pteridosperms and some conifers. It thus represents a transition between the typical Upper Carboniferous assemblages, and the Lower Permian assemblages found mainly in red-beds, and representing drier conditions (see section 5.1.1.3).

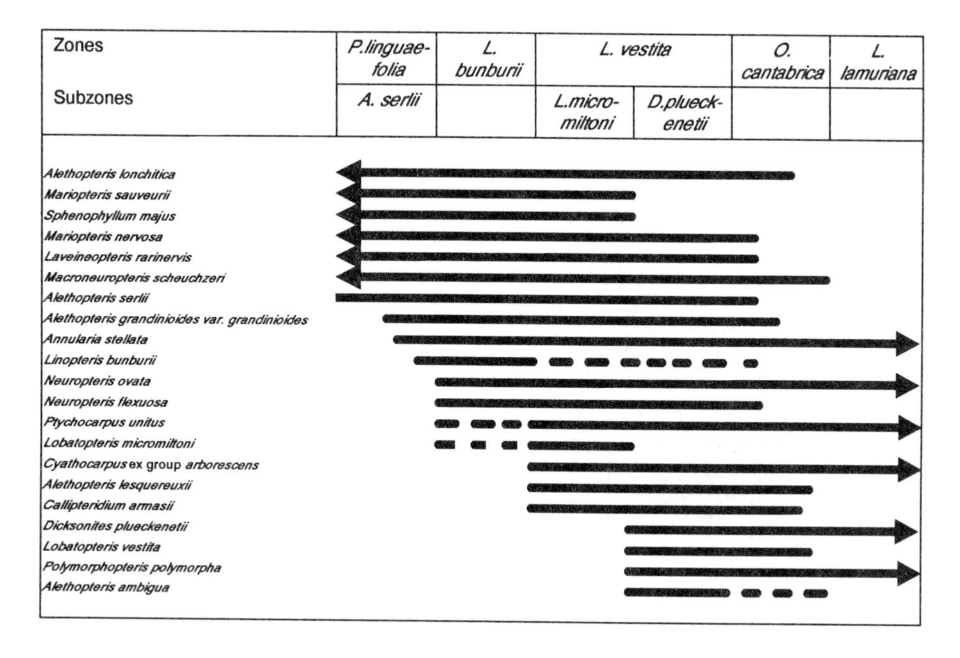

Fig. 5.6 — Stratigraphic distribution of key plant fossils in the upper Westphalian to lower Stephanian of the Europe Palaeoarea. See text for main sources of data.

Fig. 5.1 demonstrates the relationship between the biostratigraphy, summarized above, and the west European chronostratigraphy. It clearly has its lowest resolution in the Lower Carboniferous (Tournaisian–Visean) and cannot compete with the palynological zonation summarized by Owens (1984). In the Namurian and lower Westphalian, the situation improves and plant macrofossils produce as good (and sometimes better) results as the pollen and spores. Non-marine bivalves provide a significantly better resolution in the lower Westphalian of western Europe (e.g. Eagar 1956), but it is not clear over how wide a geographical area such correlations are possible with these shells. In the upper Westphalian and Stephanian, Wagner's

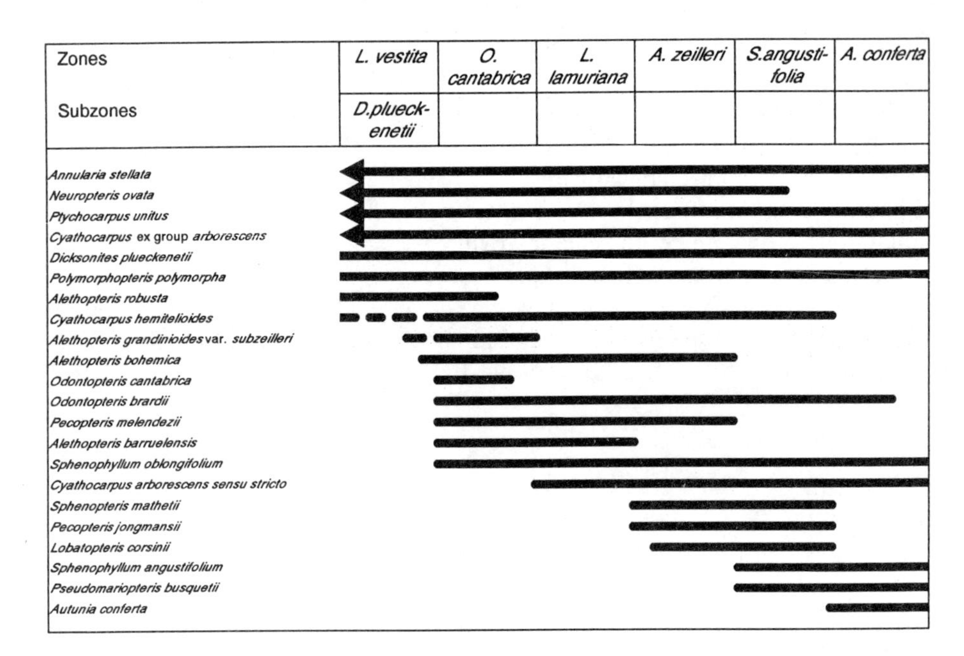

Fig. 5.7 — Stratigraphic distribution of key plant fossils in the topmost Westphalian to Lower Permian of the Europe Palaeoarea. See text for main sources of data.

biozones stand clear as the best means of establishing stratigraphic correlations in non-marine strata. For reviews of the various biostratigraphic tools available in the Carboniferous, in both non-marine and marine facies, the reader is directed to the series of papers in Sutherland & Manger (1984).

Why plant fossils have proved more useful biostratigraphically in the Upper Carboniferous than any other part of the geological column is far from clear. Wagner (1984) argued that their value resulted from a combination of rapid evolution of the floras and climatic changes, but it is not certain that either factor was more prevalent in the Late Carboniferous than at other times. It may be significant that the majority of taxa used in the Wagner Classification represent riparian plants, growing on unstable levee-banks, and would thus be subject to significant localized extrinsic stress (*sensu* DiMichele *et al.* 1987). This may have induced greater evolutionary change than, for instance, the plants growing in the more swampy palaeoequatorial habitats in the Late Carboniferous (see Phillips 1980). Alternatively, however, it may merely reflect an historical scientific bias: detailed plant fossil biostratigraphy was first attempted in the Upper Carboniferous, and this generated subsequent refinements. Perhaps further investigations in other parts of the geological column may eventually produce as good results.

5.1.1.2 Coal balls

Upper Carboniferous plant biostratigraphy in the Eurameria Palaeoarea has tended to rely almost exclusively on the adpressions, as summarized in the previous section

(section 5.1.1.1). However, these fossils represent only a small part of the vegetation of the palaeoequatorial swamp forests of the time, probably that growing on the raised levees of the fluvial distributary channels. The fragments of the plants growing in most of the forest were rarely able to find their way into the sediment-accumulating bays within the fluvio-deltaic complexes.

The plants forming the main part of the forest are best represented by the coal seams. In most situations, it is not possible to ascertain details of the plants preserved in the coals, other than in a broad sense by examining the palynology. In exceptional circumstances, however, calcareous nodules known as coal balls formed in the peat before any significant compression occurred, and these allow details of the plants to be determined. Mostly, they appear to have represented a quite distinct flora from that growing on the levees and other clastic substrates, being adapted to conditions poor in oxygen and nutrients, of low pH, and subject to periodic flooding (DiMichele *et al.* 1985).

Until recently, relatively little was known of the detailed biostratigraphic distribution of the coal ball fossils, but the situation changed significantly in the 1980s, mainly through the efforts of Phillips and his co-workers (Phillips 1980, 1981, Phillips *et al.* 1985, DiMichele *et al.* 1985). We are still far from being able to establish a formal biozonation for these fossils, and the difficulties involved in their preparation and identification mean that they will probably rarely be of significant use for establishing detailed stratigraphic correlations. Nevertheless, there are patterns in their stratigraphic distribution which are worth mentioning in the context of this chapter, especially with respect to the patterns observed in the adpression record.

1. First wet interval (Langsettian Stage). Assemblages dominated by lycophytes adapted to wetter conditions, such as *Lepidophloios harcourtii* and *Diaphorodendron vasculare*. Also typifying this interval is the lagenostomalean pteridosperm *Lyginopteris*.
2. First dry interval (Duckmantian and Bolsovian stages). The lycophytes *L. harcourtii* and *D. vasculare* decline, the former mainly in the upper part of the interval. They are partially replaced by *Paralycopodites*. Cordaite remains (*Mesoxylon*/*Mitrospermum*) form major components of some assemblages. *Lyginopteris* becomes extinct at the base of this interval, but there is a gradual increase in abundance in trigonocarpalean pteridosperms (*Medullosa, Pachytesta*) and marattialean ferns (*Psaronius*).
3. Second wet interval (Westphalian D Stage). Lycophytes favouring wet conditions again become dominant, but there are different species compared with the first wet interval: *Lepidophloios hallii, Diaphorodendron scleroticum, D. phillipsii* and *Synchysidendron dicentricum. Paralycopodites* declines through this interval. The *Mesoxylon*/*Mitrospermum* cordaites become extinct at the base of the interval, and are replaced by *Pennsylvanioxylon*/*Cardiocarpus*, which are thought to be mangrove-like trees. The marattialean ferns progressively increase in importance. Callistophyte pteridosperms also first appear in the middle of this interval.
4. Second dry interval (Barruelian and Stephanian B stages). The exact chronostratigraphic position of the base of this interval is not clear, since little data is

available from coals unequivocally belonging to the Cantabrian Stage. However, by the Barruelian a marked change in the coal-ball plant fossils has occurred. Most of the lycophytes become extinct. Exceptions include *Sigillaria* and *Polysporia*, but they never form major components. Instead, most assemblages are dominated by marattialean ferns (mostly *Psaronius*), with subsidiary trigonocarpalean and callistophyte pteridosperms, and sphenophytes (*Sphenophyllum, Arthropitys*).

5. Third wet interval (Stephanian C and Autunian stages). The existence of this brief return of wetter conditions, prior to the final aridification characterizing most of the Permian of the Eurameria Palaeokingdom, is based mainly on evidence of the distribution of coal seams (Phillips *et al.* 1985). Little is known of the composition of the coals in this interval, however, but they seem to have been formed by similar plants present in the second dry interval (marattialean ferns and subsidiary pteridosperms and sphenophytes).

Few plants grew in both the coal-forming swamps and the clastic-substrate environments, and so the coal-ball and adpression assemblages are of radically different composition. Only during drier intervals did some of the species favouring the clastic substrates also migrate into the swamps. It is therefore impossible to correlate the coal-ball species distributions in detail with the Wagner Classification for the adpressions. However, there are two points of correlation which are worth bringing out. Firstly, the extinction of *Lyginopteris* pteridosperms in the coal balls at the end of the first wet interval correlates broadly with the extinction of adpressions of that group just above the *Lyginopteris hoeninghausii* Biozone. Secondly, the appearance of callistophyte pteridosperms in coal balls in the middle of the second wet interval appears to be at about the same level as the introduction of *Dicksonites* adpression foliage, in the middle *Lobatopteris vestita* Biozone. Further correlations may eventually prove possible, but will require far more information on the relationship between the coal-ball and adpression taxa, and on the distribution of foliage types in the coal-balls.

5.1.1.3 Lower Permian adpressions

In parts of the Europe Palaeoarea, the *A. conferta* Biozone extends into the the Lower Permian, up to the basal Artinskian Stage (e.g. Doubinger 1956, Barthel 1976, Kerp & Fichter 1985). Kozur (1978) attempted to subdivide the zone, using evidence of callipterid distribution culled from the European and North American literature, but this scheme was severely criticized by Kerp (1988) as being at least partly based on doubtful palaeobotanical evidence. Haubold (1980) proposed an alternative scheme, based exclusively on evidence from Thuringia in Germany, in which the biozone was subdivided into five informal intervals (termed A_t–E_t) based mainly on the distribution of *Callipteris* sensu lato and conifers. As pointed out by Haubold, the limited amount of evidence on plant fossil distribution in other Lower Permian sequences makes it difficult to assess the usefulness of these divisions, and how far they are controlled by local palaeoecological factors. However, data from Saarland in Germany (e.g. Kerp & Fichter 1985) may eventually provide the key to establishing a more rigorous biostratigraphy for these strata. For a more detailed discussion on this topic, see Kerp (1988).

Elsewhere in the Lower Permian red-beds of this palaeoarea, plant fossils are generally sparse and, where they do occur, are predominantly conifers. Conifer remains in fact occur rarely as low as the middle Westphalian (Scott & Chaloner 1983), with further sporadic records throughout the rest of the Carboniferous (Lyons & Darrah 1989). These Carboniferous records represent the remains of extra-basinal vegetation, preserved either as small fragments that had undergone long-distance transport, or in small pockets of extra-basinal sediment that fortuitously avoided subsequent erosion. During the very late Carboniferous and Permian, however, climatic aridification caused the conifers to migrate into the lowland, sediment-accumulating habitats, with the result that conifer fossils become more widely distributed.

Despite the long stratigraphic range of these conifer fossils, it is difficult at present to determine any biostratigraphic pattern to their distribution. The problem is partly due to the difficulty of identifying conifer foliage accurately, unless they yield cuticles or are associated with fructifications. The majority of the records refer to *Walchia piniformis*, but this name has often been used for any Upper Carboniferous or Permian conifer foliage fragment. Despite the work of Florin (1938–1945), Clement-Westerhof (1984) and Visscher *et al.* (1986), considerably more information on the taxonomy and distribution of the Upper Palaeozoic conifer fossils will be needed before any practical biostratigraphy can be formulated for them.

5.1.1.4 Upper Permian adpressions

The only Upper Permian plant fossils from the Europe Palaeoarea are those from the so-called Zechstein deposits (Stoneley 1958, Schweitzer 1986). They represent a mixture of vegetation types growing near the banks of the Zechstein inland sea. Dominant are conifers (*Ullmannia, Quadrocladus, Pseudovoltzia*), but there are also rare sphenophytes (*Neocalamites*), pteridosperms (*Sphenopteris, Lepidopteris*), cycads (*Taeniopteris, Pseudoctenis*) and a possible ginkgophyte (*Sphenobaiera*). The whole aspect of the assemblage is quite different from any others from the Palaeozoic of Europe, being essentially Mesophytic in character.

Most of the plant fossils originate from the Lower Zechstein (e.g. the Kupferschiefer of Germany, the Marl Slate of Britain) but there are also records from the Upper Zechstein. However, the only difference is that the Lower Zechstein assemblages are significantly more diverse.

5.1.2 Gondwana Palaeoarea

In the Lower Carboniferous, the Gondwana plant fossils belong to the Eurameria Palaeokingdom (see section 4.2.3.2). However, assemblages are generally of restricted composition, making it difficult to apply the Wagner Classification in detail. The most complete evidence comes from eastern Australia, as reviewed by Morris (in Roberts 1985), where three zones are recognized.

1. *Leptophloeum* Biozone. This consists of extremely impoverished assemblages, consisting mainly of *Leptophloeum* and *Archaeocalamites*. Its quoted occurrence

in the lower Hastarian Stage would make it a correlative of the basal *Adiantites* Biozone in the Wagner Classification.

2. *Lepidodendropsis* Biozone. These are also extremely impoverished assemblages, this time dominated by *Lepidodendropsis*. It is stated to occur in the upper Hastarian Stage, and thus appears to correlate with part of the upper *Adiantites* Biozone of Wagner. It should be noted that in the palaeoequatorial belt *Lepidodendropsis* ranges rather higher than in Australia, probably up to the Chadian Stage.
3. *Lepidodendron* Biozone. These assemblages are dominated by arborescent lycophytes such as *Lepidodendron* and *Sublepidodendron*, together with some pteridosperm or fern fronds such as *Rhodeopteridium, Sphenopteridium, Adiantites and Cardiopteridium*. Morris states that it extends from the Ivorian to Asbian stages. Most of the pteridosperm taxa would confirm an approximate correlation with Wagner's *Triphyllopteris* Biozone, although the presence of *Cardiopteridium* suggests that it might also extend into the *N. antecedens* Biozone.

Evidence from South America, mainly Argentina, is reviewed by Rocha Campos & Archangelsky (1985). They assign all of the known (?)Lower Carboniferous assemblages to the *Archaeosigillaria* Biozone. This may correspond, at least in part, with the *Triphyllopteris* Biozone of the Wagner Classification, although Archangelsky & Azcuy (1985) have argued that some of the fossil-bearing horizons may be lower Namurian, which is a rather higher chronostratigraphic position than is usually associated with this biozone.

Plant fossils from horizons overlying the *Lepidodendron/Archaeosigillaria* Biozone have also been assigned to the Lower Carboniferous (e.g. Raymond 1985). However, it is now generally accepted that these assemblages are Upper Carboniferous, and so will be dealt with further below (section 5.4).

5.1.3 Kazakhstania Palaeoarea

The plant fossils from this palaeoarea are reviewed by Radchenko (1961, 1967, 1979, 1985). To date, no formal biostratigraphy has been established here, nor has there been any attempt to utilize any of the classifications developed elsewhere, such as that of Wagner. However, Radchenko has recognized four 'floras', which are essentially biozonal in nature.

1. 'Lower Tournaisian flora'. Impoverished and poorly preserved assemblages mainly of lycophytes. By inference, this probably equates partly with the *Adiantites* Biozone of Wagner.
2. 'Lower Visean flora'. Also poorly preserved assemblages, but including *Lepidodendropsis vandergrachtii, Lepidodendron volkmannianum, Archaeocalamites radiatus, Sphenophyllum tenerrimum* and *Neurocardiopteris* sp. This appears to correspond to the *Triphyllopteris* Biozone of Wagner.
3. 'Visean–Namurian flora'. Diverse assemblages including *A. radiatus, S. tenerrimum, Mesocalamites* spp., *Sphenopteris bifida, Lyginopteris bermudensiformis,*

L. fragilis and *Neuropteris antecedens*. This appears to correspond to the *Neuropteris antecedens* Biozone of Wagner.

4. 'Bashkirian flora'. Assemblages dominated by endemic taxa, but also including *Mesocalamites* spp., *Calamites* spp., *Alloiopteris coralloides* and *Renaultia gracilis*. The decline in number of cosmopolitan taxa makes it difficult to place this in the Wagner Classification, but has most in common with the *Lyginopteris larischii* Biozone.

A detailed account of assemblages representing each of the above biozones is given by Radchenko (1985). Assemblages from higher horizons take on an Angaran aspect, and so are reviewed below (section 5.5.2).

5.1.4 Cathaysia Palaeoarea

Plant fossils from the Carboniferous of the Cathaysia palaeocontinent clearly belong to the Eurameria Palaeokingdom; only in the Permian is a separate palaeokingdom recognized (see sections 4.2.3.4 and 4.5). In principle, therefore, it should be possible to apply at least the biozones of the Wagner Classification to these earlier assemblages. As pointed out by Laveine *et al.* (1989), however, there are discrepancies in the stratigraphic ranges of some of the taxa between Cathaysia and Europe, which makes a detailed application of the Wagner Classification to the former palaeoarea difficult.

The plant fossil biostratigraphy of the Cathaysia palaeocontinent is reviewed by Yang *et al.* (1983), Li & Yao (1985), Zhao & Wu (1985) and Zhang (1987). The following biozones are now generally recognized, with minor modifications to the nomenclature for the sake of simplicity.

1. *Sublepidodendron mirabile* Biozone. Assemblages including *S. mirabile*, *Lepidodendropsis hirmeri* and *Archaeocalamites* sp. There are many similarities with assemblages from the Upper Devonian, but they lack typically Devonian elements such as *Cyclostigma kiltorkense*, *Leptophloeum rhombicum* and *Archaeopteris macilenta*. The available evidence suggests that this biozone occurs in the upper Tournaisian and is thus approximately coeval with the lower *Triphyllopteris* Biozone in the Wagner Classification.
2. *Triphyllopteris collumbiana* Biozone. This includes some of the commonest of the Lower Carboniferous assemblages from China, characterized by species such as *T. collumbiana*, *Cardiopteridium spetsbergense*, *Lepidodendron volkmannianum*, *L. rhodeanum*, *Archaeocalamites radiatus*, *Sphenophyllum tenerrimum*, *Paripteris gigantea* and *Neuropteris* cf. *antecedens*. It occurs in the Datangian Stage in the Chinese chronostratigraphic classification (broadly equivalent to the Visean), and is probably equivalent to the upper *Triphyllopteris* and lower *N. antecedens* biozones in the Wagner Classification. It is interesting to note, however, the much lower occurrence here of *Paripteris gigantea*, compared with in Europe (Laveine *et al.* 1989). There is evidence of a gradual transition between this and the underlying *S. mirabile* Biozone, such as in the Gaolishan Formation of the Lower Yangtze.

3. *Eleutherophyllum mirabile* Biozone. This only occurs in a few localities in China. Characteristic species include *E. mirabile, Mesocalamites cistiformis, Paripteris gigantea and Linopteris neuropteroides*. They are often associated with Arnsbergian goniatites, which suggests that they are coeval with either the *L. stangeri* or *L. larischii* biozones of the Wagner Classification.
4. *Lepidodendron aolungpylukense* Biozone. This is mainly restricted to the Qilian Mountains and adjacent areas. Most of the reported species are Cathaysian endemics, but also include *Lepidodendron volkmannianum* and *Karinopteris acuta*. It is coeval with part of the *P. aspera* Biozone of the Wagner Classification.
5. *Paripteris pseudogigantea* Biozone. This is the so-called Penchi Formation 'flora', found extensively in north China. Important species include *Lepidodendron worthenii, Paripteris pseudogigantea, Linopteris brongniartii, L. neuropteroides and Eusphenopteris obtusiloba*. Yang *et al*. (1983) record the assemblage as spanning the Dalan Stage in the Cathaysia chronostratigraphy, which is regarded as broadly equivalent to the Westphalian Series in Europe. From the species quoted, it is tempting to equate it with the lower and middle Westphalian biozones in the Wagner Classification (*L. hoeninghausii, L. rugosa* and *P. linguaefolia* biozones), but the problems inherent in intercontinental correlations using plant fossils (Laveine *et al*. 1989) must make this suggestion very tentative.
6. *Lepidodendron posthumii* Biozone. This is mainly represented in the Taiyuan Formation of North China and the Djambi 'flora' of Sumatra. The general composition of the biozone shows similarities with coeval assemblages in the Eurameria Palaeokingdom, with abundant trigonocarpalean pteridosperms (*Neuropteris, Alethopteris, Callipteridium*) marattialean ferns (*Cyathocarpus*) and bowmanitalean sphenophytes (*Sphenophyllum*). Unlike the Eurameria assemblages, however, lycophytes do not decline in importance, and there are abundant *Lepidodendron, Bothrodendron* and *Cathaysiodendron*. It may correlate with the *L. bunburii, L. vestita* and all of the Stephanian biozones in the Europe Palaeoarea.

It should be noted that this scheme differs slightly from that of some recent authors (e.g. Zhang 1987), particularly in the biozonation of the lower Namurian (the Huashibanian Stage of the Chinese chronostratigraphy). In the classification outlined by Zhang, the *E. mirabile* and *L. aolungpylukense* biozones are not separated, the combined unit being referred to as the *E. mirabile–Linopteris* sp.–*L. aolungpylukense* Biozone in north China, and the *Paripteris gigantea–Karinopteris acuta* forma *obtusa* Biozone in south China.

Plant fossils from above the *L. posthumii* Biozone take on an exclusively Cathaysian aspect, and so their biostratigraphy is dealt with later (section 5.3).

5.2 NORTH AMERICA PALAEOKINGDOM

This palaeokingdom only becomes identifiable in the Leonardian Series (Lower Permian). Older assemblages may be assignable to a separate palaeoarea (the Cordillera Palaeoarea; see section 4.2.3.5) but they are sufficiently similar to the rest of the Eurameria Palaeokingdom to allow the biostratigraphies of Read & Mamay (1964) and Wagner (1984) to be applied.

The most significant biostratigraphic analysis of these assemblages is by Read & Mamay (1964), who recognize two biozones: 'Zone of the older *Gigantopteris* flora' and 'Zone of the younger *Gigantopteris* flora'. Subsequent work has resulted in many of the taxa being re-named (e.g. Mamay 1986, 1989, Mamay *et al.* 1988), and so a revised nomenclature for these biozones is proposed here.

1. *Gigantopteridium americanum* Biozone (lower Leonardian), defined by the range of *G. americanum*. Also restricted to it are *Russellites taeniata, Glenopteris* spp. and *Supaia* spp., although they are only locally abundant.
2. *Cathaysiopteris yochelsonii* Biozone (upper Leonardian), defined by the occurrence of *C. yochelsonii, Zeilleropteris wattii, Evolsonia texana, Delnortia abbottiae* and possibly *Tinsleya texana* (although see comments by Kerp (1988) on its possible synonymy with *Rhachiphyllum schenkii*).

Read & Mamay (1964) subdivided what is called here the *G. americanum* Biozone into '*Gigantopteris*', *Supaia* and *Glenopteris* 'floras'. However, problems of detailed correlation make it uncertain whether this distinction is biostratigraphic, palaeogeographic or palaeoecological in nature.

5.3 CATHAYSIA PALAEOKINGDOM

In the Permian, Cathaysian plant fossil assemblages have developed into a discrete palaeokingdom, and a quite separate biostratigraphy has been developed to describe their temporal change. There are a number of reviews of the subject, the most significant by Gu & Zhi (1974), Yang *et al.* (1983), Mei (1984), Li & Yao (1985), Zhao & Wu (1985) and Zhang (1987).

The currently accepted classification is based on Li (1963, 1964), whose work was based primarily on the North China Palaeoarea. However, the scheme has subsequently been extended into the South China Palaeoarea, principally by Li & Yao (1985). Four biozones are recognized in this scheme, which are shown in Figs 5.8–5.10, together with the ranges of some of the taxa used to identify the boundaries. The *E. alatum* and *C. whitei* biozones are regarded as Lower Permian, and the *Otofolium* and *Ullmannia* biozones as Upper Permian. However, there is no detailed correlation yet available between this biostratigraphy and any of the generally accepted chronostratigraphies.

The *E. alatum* Biozone is widely distributed in Cathaysia, being found in North China (e.g. the Shanxi Formation), South China (e.g. the Lungtan Formation) and Sumatra (the upper Djambi 'flora'). It marks the start of the fully developed Cathaysian assemblages, with practically no Euramerian elements being present. Lycophytes, although somewhat less diverse than in the *L. posthumii* Biozone, continue to be important components. Sphenophytes and ferns also remain important. Of particular significance is the appearance of tripinnate fronds such as *Emplectopteris* and *Emplectopteridium*, which are thought to be the precursors of the gigantopterid leaves that characterize the Upper Permian Cathaysian assemblages (Assama 1959).

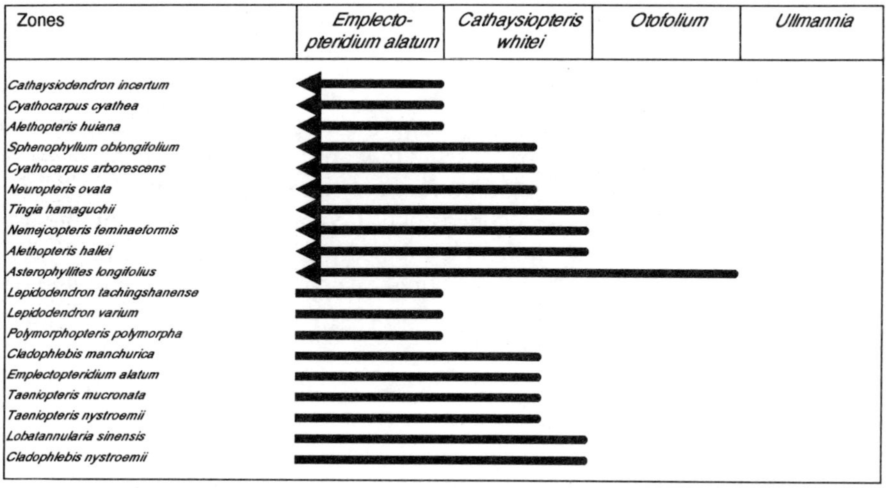

Zones | *Emplecto-pteridium alatum* | *Cathaysiopteris whitei* | *Otofolium* | *Ullmannia*

Cathaysiodendron incertum
Cyathocarpus cyathea
Alethopteris huiana
Sphenophyllum oblongifolium
Cyathocarpus arborescens
Neuropteris ovata
Tingia hamaguchii
Nemejcopteris feminaeformis
Alethopteris hallei
Asterophyllites longifolius
Lepidodendron tachingshanense
Lepidodendron varium
Polymorphopteris polymorpha
Cladophlebis manchurica
Emplectopteridium alatum
Taeniopteris mucronata
Taeniopteris nystroemii
Lobatannularia sinensis
Cladophlebis nystroemii

Fig. 5.8 — Stratigraphic distribution of key plant fossils in the Permian of the Cathaysia Palaeokingdom. See text for main sources of data.

Zones | *Emplecto-pteridium alatum* | *Cathaysiopteris whitei* | *Otofolium* | *Ullmannia*

Emplectopteris triangularis
Taeniopteris multinervis
Polymorphopteris wongii
Taeniopteris serrulata
Annularia mucronata
Plagiozamites oblongifolius
Sphenopteris norinii
Lepidodendron cervicisum
Sphenophyllum neofimbriatum
Cathaysiopteris whitei
Taeniopteris shansiensis
Pterophyllum cutelliforme
Pecopteris hirta
Gigantoclea lagrellii
Sphenophyllum costae
Tingia crassinervis
Sphenopteris gothanii
Lepidodendron polygonum
Sphenophyllum sinocoreanum

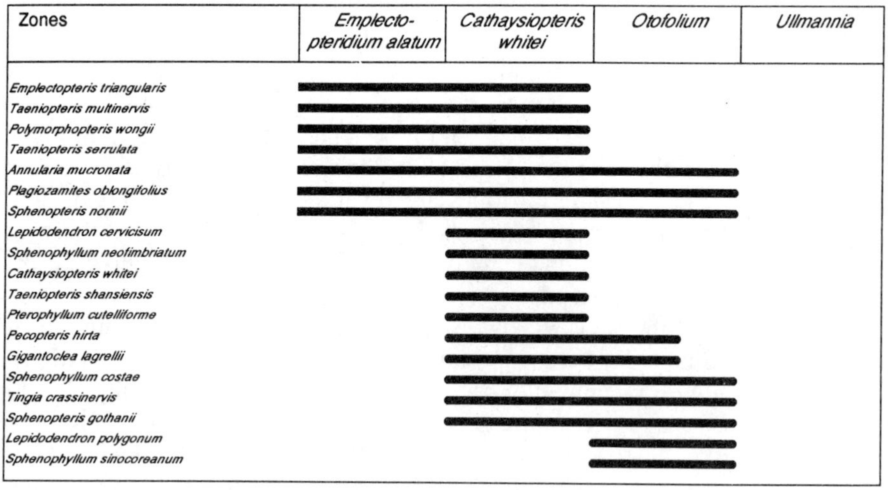

Fig. 5.9 — Stratigraphic distribution of key plant fossils in the Permian of the Cathaysia Palaeokingdom. See text for main sources of data.

Zones	*Emplecto-pteridium alatum*	*Cathaysiopteris whitei*	*Otofolium*	*Ullmannia*
Sphenophyllum speciosum			▬	
Sphenophyllum koboense			▬	
Yuania striata			▬	
Pecopteris echinata			▬	
Gigantoclea acuminatiloba			▬	
Taeniopteris taiyuanensis			▬	
Taeniopteris serrata			▬	
Pterophyllum eratum			▬	
Walchia bipinnata			▬	
Otofolium polymorphum			▬	
Annularia shirakii			▬	▬
Lobatannularia multifolia			▬	▬
Schizoneura manchuriensis			▬	▬
Gigantopteris dictyophylloides			▬	▬
Gigantopteris nicotianaefolia			▬	▬
Neuropteridium coreanicum			▬	▬
Rhipidopsis			▬	▬
Psygmophyllum multipartitum			▬	▬
Ullmannia cf. *bronnii*				▬

Fig. 5.10 — Stratigraphic distribution of key plant fossils in the Permian of the Cathaysia Palaeokingdom. See text for main sources of data.

The *C. whitei* Biozone occurs typically in the Upper Shihhotse Formation of North China and the Lungtan Formation of South China. Lycophytes, sphenophytes and ferns remain common here. The gigantopteroid fronds show a trend towards a lower level of pinnation, with bipinnate and monopinnate forms such as *Cathaysiopteris* and *Gigantonoclea*. Also of interest is the appearance of early cycad-like fronds.

The Upper Shihhotse Formation yields the best known of the *Otofolium* Biozone assemblages, although they are also found extensively in the Lungtan Formation of South China. The earliest true gigantopterids with undivided leaves appear here, these being regarded by many as the most characteristic plant fossils of the Cathaysia Palaeokingdom (although very similar leaves are also known from the Lower Permian of North America; e.g. Mamay 1986, 1989). Lycophytes undergo a marked decline, but sphenophytes and ferns continue to be abundant. There are also several species of cycad-like leaves (*Taeniopteris*).

The *Ullmannia* Biozone marks a significant decline in the number and diversity of Cathaysian assemblages. The few known examples include those reported from the Linchai Formation of Liaoning (where there appears to be a mixture of Cathaysian and Angaran elements) and the Jinjibang Formation of Sichuan. Most ferns, lycophytes and sphenophytes become extinct, whilst conifers become more important. Simple gigantopterid leaves are also common, and their precursors with divided leaves (e.g. *Gigantonoclea*) have disappeared. The zone may be taken as the start of the gradual transition between Palaeophytic and Mesophytic 'floras', which takes place at a significantly higher stratigraphic level than in the Eurameria Palaeokingdom.

In North China, strata overlying the *Ullmannia* Biozone (the Liujiagou Formation) is regarded as basal Triassic. It contains a limited range of plant fossils indicating the *Pleuromeia* Biozone (e.g. *Pleuromeia*, *Willsiostrobus*, *Macrostachya*,

Peltaspermum, Neocalamites). Wang (1989) has analysed the changes in plant fossil assemblages statistically over the Permian–Triassic boundary in this part of China, and concluded that there was a major extinction event at this level. This is in contrast to the results obtained in other parts of the world (e.g. Boulter *et al.* 1988), where the major change appears to occur in the mid-Permian (the so-called Palaeophytic–Mesophytic transition).

To date, there has been no concerted effort to establish a palynological biozonation for the Permian of the Cathaysia palaeocontinent (Gao Zhi-feng, personal communication). Plant fossils are thus the only means available at present of correlating the non-marine strata of this age in this part of the world.

5.4 GONDWANA PALAEOKINGDOM

There have been a number of reviews of Gondwanan plant biostratigraphy, the most helpful by Surange (1975), Schopf & Askin (1980), Archangelsky *et al.* (1980), Archangelsky (1984), Archangelsky & Azcuy (1985), Rocha Campos & Archangelsky (1985), Rigby (1985), Roberts (1985), Tripathi & Singh (1985) and Anderson & Anderson (1985). Most have adopted an essentially classical methodology, attempting to recognize the first and last appearances of key taxa. In contrast, Retallack (1980) has tried to use the principles of ecostratigraphy to produce more refined results (see also Retallack 1978). He established a series of associations or 'florules', whose distribution was thought to be controlled by environmental factors as much as genetic evolution. This approach undoubtedly holds out much promise for improving the resolution of correlations using plant fossils. As pointed out by Rigby (1985), however, our present understanding of the environmental controls on these plants is still too meagre for ecostratigraphy to play a major role in establishing long-distance stratigraphic correlations.

Throughout the palaeokingdom, there is a uniformity in the broad stratigraphic distribution of Upper Carboniferous and Permian plant fossils. In essence, it may be summarized as a mainly Upper Carboniferous 'pre-*Glossopteris*' interval succeeded by a mainly Permian *Glossopteris–Gangamopteris* interval. In various parts of Gondwana, there have been differences in detail in the biostratigraphic schemes used. However, this is at least partially due to subjective differences in approach adopted by authors working in different countries. In principle, it should be possible to establish a unified scheme for use throughout the Gondwana Palaeokingdom, similar to the Wagner Classification for the Eurameria Palaeokingdom. The outlines of such a scheme are summarized in Fig. 5.11.

The *N. argentinica* Biozone ranges from upper Visean to mid-Westphalian. It refers to relatively impoverished assemblages dominated by *Nothorhacopteris argentinica, Dichophyllites peruvianus*, '*Lepidodendropsis*' *peruviana* and '*Cyclostigma*' *pacificum* (see Archangelsky *et al.* 1981 for comments on the generic position of these latter two species). In the Nothoafroamerica Palaeoarea, there are also a number of endemic lycophytes such as *Bumbodendron* and *Brasilodendron*. It is widely distributed, with records from the Po Formation in India, the Cortadera and Tupe formations in Argentina, The Ambò Formation in Peru and the McInnes and Italia Road formations of Australia.

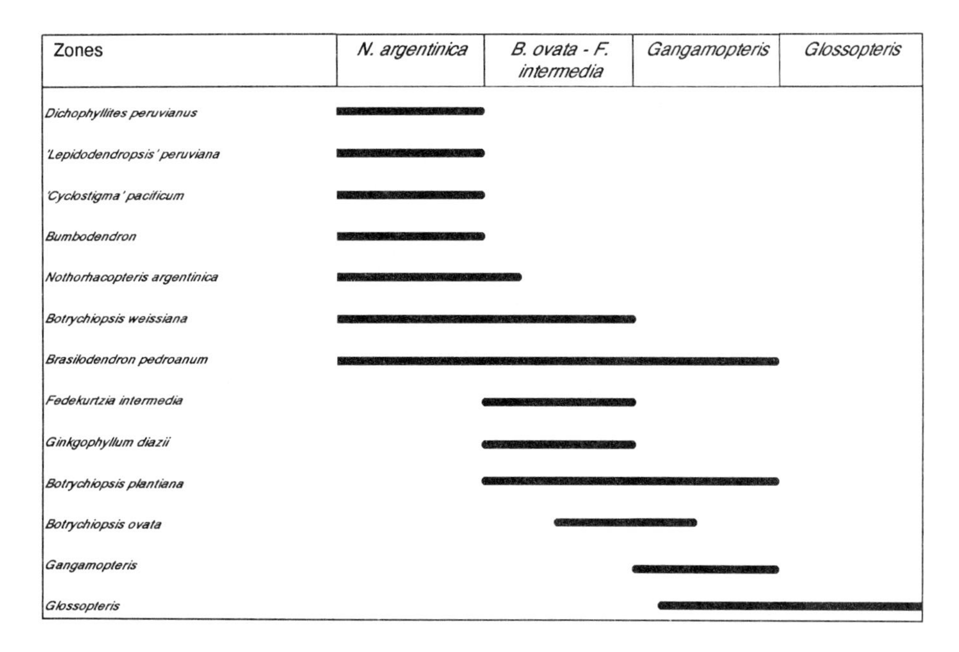

Fig. 5.11 — Stratigraphic distribution of key plant fossils in the Upper Carboniferous and Permian of the Gondwana Palaeokingdom. See text for main sources of data.

The succeeding *B. ovata–F. intermedia* Biozone retains many of the characteristics of the *N. argentinica* Biozone. *N. argentinica* continues up into this biozone, but also present are abundant *B. ovata* and *F. intermedia*. The two latter species rarely occur together, owing to ecological differences, and Retallack (1980) recognized two separate 'florules'. However, they appear to overlap in time and so are united here as a single biozone. The most complete records are from Australia, such as from the Joe Joe Group, the Currabubula Formation and the top of the McInnes and Italia Road formations. It is also probably represented in the Malanzán Formation and its lateral equivalents in Argentina, which Rocha Campos & Archangelsky (1985) refer to as the *Ginkgophyllum* Biozone. It ranges from mid-Westphalian to upper Stephanian.

The *Gangamopteris* Biozone represents the sudden influx of 'glossopterid' (Arberiales) plant fossils in the topmost Carboniferous or basal Permian (the exact chronostratigraphic position of the base of this biozone is still a matter of dispute). Although the biozone contains a variety of distinctive taxa, including conifers, ferns and sphenophytes, it is most readily identified by the presence together of *Glossopteris* and *Gangamopteris*, which almost invariably occur and often dominate the assemblages. Assemblages assignable to this biozone are known from the Ecca Group in South Africa, the Talchir, Karharbi and possibly the basal Barakar groups in India, the Lubeck and lower Golondrin formations in Argentina, and from the Lower Permian of Australia and Antarctica.

In the Upper Permian, *Gangamopteris* becomes extinct and the interval is thus recognized as a discrete *Glossopteris* Biozone. The base of the biozone corresponds approximately with the boundary between the Lower and Upper Permian, although its exact chronostratigraphic position is still a matter of conjecture. The assemblages tend to consist almost exclusively of *Glossopteris*, with only rare marattialean ferns (*Asterotheca auct.*) and sphenophytes (*Sphenophyllum*). They are best documented from the topmost Ecca Group and Beaufort Group of South Africa, the Raniganj and most (if not all) of the Barakar groups of India, and the upper Golondrin Formation of Argentina.

There have been attempts at further subdivision of the *Gangamopteris* and *Glossopteris* biozones. In India, five 'floral stages' have been recognized, based partly on differences in the *Glossopteris* species present (Chandra & Surange 1979). In the Paraná Basin in Brazil, Rössler (1978) distinguished 'Taphofloras' B and C, based on differences in the species of *Glossopteris* and *Gangamopteris*, and the absence of sphenophyte foliage *Annularia* from the upper unit (C). In Patagonia, Archangelsky & Cúneo (1984) recognize, in ascending order, *Nothorhacopteris chubutiana, Ginkgoites exima, Dizeugotheca waltonii* and *Asterotheca singeri* biozones, based on clear-cut differences in the assemblages (see Archangelsky & Cúneo 1984, Table 2 for details). However, these schemes are essentially of local significance and will not be discussed further here.

There is as yet no palynological biozonation that has been proposed for use throughout the Gondwana palaeocontinent and could be compared with the above plant fossil biozonation. In most areas, large monosaccate pollen dominate the Upper Carboniferous assemblages, and taeniate bisaccate pollen the Permian. This effectively corresponds to the boundary between the *B. ovata–F. intermedia* and the *Gangamopteris* biozones, although the palynological change is gradational. In some of the areas within the Gondwana palaeocontinent, however, far more refined palynological biostratigraphies have been proposed, such as by Daemon & Quadros (1970) for Brazil, Anderson (1977) for South Africa, and Kemp *et al.* (1977) for Australia (see Schopf & Askin 1980 for a more complete review). These clearly are the same type of local biostratigraphy as has been proposed for some of these areas based on the plant fossils (see previous paragraph), and with which any qualitative comparison should be made.

5.5 ANGARA PALAEOKINGDOM

Plant fossils have been extensively used for biostratigraphic work in this palaeokingdom. The assemblages tend to have a lower species diversity than those of the Eurameria Palaeokingdom, with the result that the stratigraphic resolution that they provide is significantly poorer. Nevertheless, in some facies they provide one of the most reliable means of making correlations. Major reviews have been provided by Neuburg (1948, 1961), Graizer (1967), Gorelova *et al.* (1973), Durante (1976), Vakhrameev *et al.* (1978), Radchenko (1979, 1985) and Meyen (1982, 1987).

5.5.1 Kuznetsk Palaeoarea

The most complete evidence of Angaran plant fossils is found in the Minussinsk and Tuva basins for the Lower Carboniferous, and the Kuznetsk Basin for the Upper

Carboniferous and Permian. Based mainly on evidence from these areas, Soviet palaeontologists have established a sequence of 'floral complexes', which partially correspond to biozones in the sense of Hedberg (1976). Unlike traditional biozones, however, they are not named after diagnostic taxa, but rather after the lithostratigraphic interval(s) in which they commonly occur. Not only does this diverge from what has become established stratigraphic usage (i.e. the use of taxonomic names for biozones), it is also a potential source of nomenclatural confusion, as the 'complexes' do not always coincide exactly with the eponymous lithostratigraphic intervals (compare columns 3 and 4 in Gorelova *et al.* 1973, plates 10–12).

Another difficulty with the 'floral complexes' is that they are not always clearly delineated by the ranges of the plant fossils. In some cases, particularly in the Upper Permian, they tend to be defined by a particular balance in the proportional composition of assemblages, which can be difficult to recognize in isolated assemblages. Nevertheless, there are a number of quite clearly delineated biostratigraphic intervals, which can be made the basis of biozones, in the sense adopted elsewhere in this chapter. The means of recognizing these are summarized in Figs 5.13–5.15, based on the range charts presented by Graizer (1967) and Gorelova *et al.* (1973). The latter studies give a far more complete account of the plant biostratigraphy of these areas than can be given here and should be consulted in conjunction with Figs 5.13–5.15. The names which are proposed here for the biozones follow, in part, the informal usage shown in Durante (1976, Tables 1–2). The correlation between the traditional Soviet 'complex' names and the new biozonal names is summarized in Fig. 5.12.

The *Lepidodendropsis* Biozone contains a mixture of Euramerian and endemic Angaran elements. The composition of the *A. alternans* and *Tomiodendron* biozones, however, is almost exclusively endemic, although Meyen (1987) compared the ferns and lycophytes with those found in the Euramerian Upper Devonian. He explains this in terms of the climatic aridity of the northern palaeocontinents in the Early Carboniferous.

The *R. yavorskyi* Biozone sees a marked decline in the proportion of lycophytes and a corresponding increase in pteridosperms; some Soviet authors have referred to it as the 'Pteridospermalean Assemblage'. It is thought to be a consequence of climatic cooling in the northern palaeocontinents (the 'Ostrogian episode'), and may be coeval with the mid-Carboniferous 'flora break' recognized in the Eurameria Palaeokingdom.

The *N. izylensis* Biozone represents the start of the 'classic' Angaran, cordaite-dominated assemblages, and coincides with the appearance of significant coal deposits in the area. It probably indicates a warming of the climate, although the presence of growth-rings in some gymnosperm woods indicates some seasonal variation. The succeeding biozones see a steady increase in the diversity of the cordaite components. There is also the introduction of a number of form-genera, normally associated with the Eurameria Palaeokingdom, which probably represent plants brought in on the Kazakhstania Palaeocontinent when it collided with the main Angara landmass in the mid-Carboniferous (see section 4.3.3.2). The evidence generally suggests a gradual climatic warming in the northern palaeocontinents during the Late Carboniferous and Early Permian.

Systems	Soviet Stages	Soviet 'Floral Complexes'	Biozones
PERMIAN	Tatarian	Tailuganskaya	*Pecopteris oviformis*
		Gramoteinskaya	
		Leninskaya	
		Uskatskaya	*Glottophyllum primaevum*
	Kazanian	Kazanovo-Markinskaya	
	Kungurian	Kuznetskaya	*Nephropsis lampadiformis*
		Usyatskaya	*Zamiopteris*
	Artinskian	Kemerovskaya	
	Sakmarian	Ishanovskaya	
	Asselian	Promezhutochnaya	*Evenkiella zamiopteroides*
CARBON-IFEROUS	Gzhelian to Moscovian	Alykaevskaya	*Phyllotheca tomiensis*
	Bashkirian	Mazurovskaya	*Neuropteris izylensis*
	Serpukhov-ian	Kaezovskaya	*Rhodeopteridium yavorskii*
		Evseevskaya	*Tomiodendron*
	Visean	III	*Angarophloios alternans*
	Tournaisian	I	*Lepidodendropsis*

Fig. 5.12 — Plant biostratigraphy for the Angara Palaeokingdom, based mainly on the distribution of adpressions in the Kuznetsk Palaeoarea.

The boundaries between the Carboniferous and Lower Permian biozones in Angara are sharply delineated and thus easily recognized. In the Upper Permian, however, they tend to be more gradational; Gorelova *et al.* (1973) give transitional zones ('Smeshanyi') between each of their 'floral complexes'. Gorelova *et al.* identify six 'floral complexes' in the Upper Permian, but distinguishing them using only the ranges of the plant fossils is difficult in many cases. An analysis of their range charts

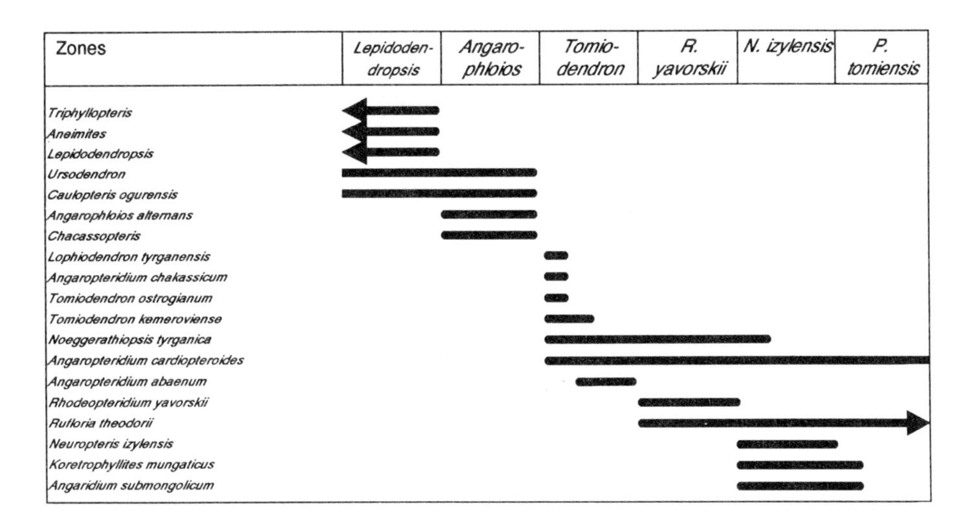

Fig. 5.13 — Stratigraphic distribution of key plant fossils in the Carboniferous of the Angara Palaeokingdom. See text for main sources of data.

(particularly their Figs 13–16) reveals just three clearly delineated biostratigraphic units, and so only these are recognized here as biozones.

The *N. lampadiformis* Biozone sees the start of the gradual introduction of more typically Mesozoic taxa, such as *Rhipidopsis*, *Tomia* and *Yavorskiya*, and corresponds broadly to the Palaeophytic–Mesophytic boundary in the Angara Palaeokingdom. The trend continues through the *G. primaevum* and *P. oviformis* biozones, and culminates in the so-called Korvunchanskaya 'flora'. The latter, which will not be dealt with in detail here, appears to straddle the Permo-Triassic boundary on the evidence of arthropod fossils, and contains a mixture of characteristically Permian and Triassic plant fossils (Meyen 1987).

Finally worth mentioning are Angaran-type assemblages from present-day north-west China. Few of these assemblages have been described in detail, but a summary of some of the taxa is provided by Zhang (1987). He divided them into two main biozones (or 'assemblages'): the *Lepidodendron–Lepidodendropsis* Biozone, reported to be Visean; and the *Angaropteridium–Mesocalamites* Biozone, from the Namurian. The former may equate with the *Tomiodendron* Biozone in the standard Angaran biostratigraphy. The latter is difficult to correlate, however, as there are either misidentifications in the quoted list, or there are major discrepancies in the species distributions in the two areas. For instance, *Chacassopteris* does not occur above the mid-*Tomiodendron* Biozone in Kuznetsk, while *Angaropteridium ligulatum* Neuburg does not range below the upper *P. tomiensis* Biozone. Clearly these Chinese assemblages need to be documented in more detail.

5.5.2 Kazakhstania Palaeoarea

Kazakhstania collided with the main Angara landmass, probably sometime in the Moscovian, and thereafter the plant fossils take on an increasingly Angaran aspect.

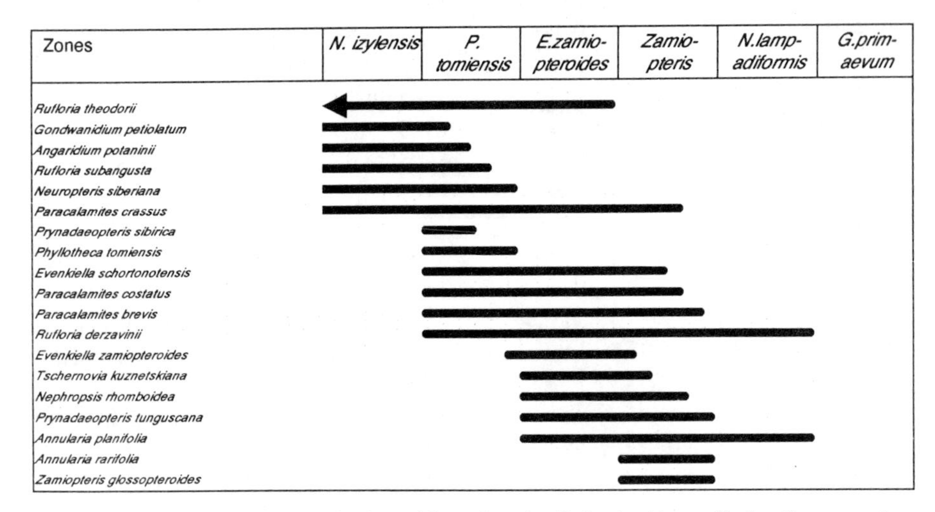

Fig. 5.14 — Stratigraphic distribution of key plant fossils in the Upper Carboniferous and Permian of the Angara Palaeokingdom. See text for main sources of data.

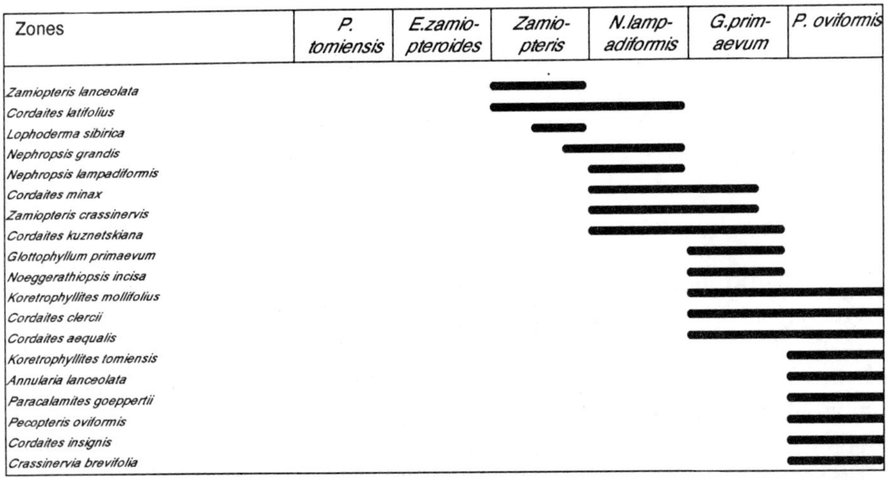

Fig. 5.15 — Stratigraphic distribution of key plant fossils in the Upper Carboniferous and Permian of the Angara Palaeokingdom. See text for main sources of data.

In eastern Kazakhstan, both the *N. izylensis* and *P. tomiensis* biozones can be recognized (Radchenko 1979, 1985). In central and southern Kazakhstan, the situation is less clear, but Radchenko gives evidence that the *P. tomiensis* Biozone is present in the Tentekskaya 'Suite' (see also Maiboroda 1979).

Lower Permian plant fossils from Kazakhstan have been poorly documented (Vakhrameev *et al.* 1978) and are thus difficult to correlate with the biostratigraphy established in the Kuznetsk Palaeoarea. The assemblages are often dominated by conifers or endemic lycophytes, quite different from those of Kuznetsk. The presence of *Zamiopteris* in north Pribalkhashie might, however, indicate the presence of the biozone of that name (Salmenova 1979).

Problems of correlation with the Kuznetsk Palaeoarea also exist in the Upper Permian. In the Kazanian and lower Tatarian, the assemblages are dominated by conifers and endemic peltasperms. They offer little comparison with the Kuznetsk assemblages but, from their chronostratigraphic position, they are probably coeval with the *Glottophyllum primaevum* Biozone. In the upper Tatarian, the strata are mainly red-beds, but occasional lacustrine deposits have yielded a distinctive assemblage, which can be referred to as the *Tatarina* Biozone (the *Tatarina* 'flora' of Gomankov & Meyen 1986). This again yields mainly endemic forms offering little in the way of direct comparison with the Kuznetsk assemblages; one of the few species in common is *Cordaites clercii*.

In conclusion, much still needs to be done in establishing detailed correlations between the Kazakhstania and Kuznetsk palaeoareas. The position in the Upper Carboniferous is not too problematic but in the Permian, only broad correlations are possible at present.

5.5.3 Pechora Palaeoarea

This palaeoarea is first recognizable in the Kungurian (the next lowest plant fossils are from the Visean, and appear to have Euramerian affinities; Vakhrameev *et al.* 1978). The most detailed information on the Pechora plant biostratigraphy is given by Meyen (1983), although, unlike the Kuznetsk, no 'floral complexes' are proposed.

The lower part of the sequence, referred to as the Vorkutsk 'Series', belongs to the Kungurian and is thus coeval with the *Zamiopteris* Biozone of the Kuznetsk Palaeoarea. That *Annulina* is virtually restricted to the Vorkutsk strata, and pteridosperms such as *Callipteris* sensu lato, *Compsopteris* and *Comia* are absent, points to some comparison with the *Zamiopteris* Biozone. However, the marked differences in the species of plant fossils makes it difficult to assign the Vorkutsk strata unequivocally to this biozone.

The overlying Pechorska 'Series' sees the introduction of a number of taxa, including *Cordaites clercii* and *Rufloria brevifolia*. This may indicate that these strata belong to the *Glottophyllum primaevum* Biozone, although, again, major differences in the species present must make this suggestion tentative. If correct, however, it indicates that the *Nephropsis lampadiformis* Biozone is absent from the Pechora Palaeoarea.

Meyen (1987) suggests that the plant fossils from the upper Pechorska 'Series' belong to the *Tatarina* Biozone, as recognized in the Kazakhstan Palaeoarea. There are, however, no species in common with the type *Tatarina* Biozone, the suggestion being based on the presence in the upper Pechorska 'Series' of form-genera such as *Tatarina* and *Aequistoma*. There also seems to be a discrepancy in chronostratigraphic position, the Kazakhstania *Tatarina* Biozone being upper Tatarian, whilst the upper Pechorska 'Series' is basal Tatarian.

5.5.4 Far East Palaeoarea

This palaeoarea is first recognizable in the Kungurian. Its plant biostratigraphy is best documented in Mongolia, through the work of Durante (1976). She essentially adopted the biozonation proposed by Meyen (1966) based on the cordaites, recognizing two 'floral complexes', referred to here as biozones.

1. *Cordaites gracilentus–Rufloria brevifolia* Biozone. According to the chronostratigraphic range of this biozone given by Durante, this would correlate with the *G. primaevum* and basal *P. oviformis* biozones of the Kuznetsk Palaeoarea. The range of *C. gracilentus* given in Gorelova *et al.* (1973) supports this view, as does the presence of *Rufloria olzerassica, Paracalamites* cf. *angustus* and *Glottophyllum karpovii*.
2. '*Cordaites* alone' Biozone. This refers to the interval in the topmost Permian where *Cordaites* persists but *Rufloria* is absent, and probably coincides with the middle and upper *P. oviformis* Biozone. The plant fossil assemblages in this interval in Mongolia are of relatively limited diversity, but the presence of *Paracalamites* cf. *goeppertii* supports this correlation.

5.6 CONCLUDING REMARKS

From the above review, it is evident that plant macrofossils can play a major role in establishing stratigraphic correlations in non-marine Carboniferous and Permian strata. For local correlations, other types of fossils can sometimes provide better resolution. Examples include the Westphalian coalfields of Britain, where non-marine bivalves can sometimes provide more refined results (e.g. Eagar 1956), and the Permian of Australia and South Africa, where palynology at present produces the best results (Kemp *et al.* 1977, Anderson 1977). Even for local correlations, however, plant macrofossils can sometimes better other types of fossils, as has been shown in the upper Westphalian and lower Stephanian of northern Europe (e.g. Cleal 1978, 1984, 1986, 1987).

Plant fossil biostratigraphy really comes into its own for what may be termed medium-range correlations, i.e. within a particular palaeoarea. At this level, plant fossils often provide as good if not better resolution than other fossils, such as pollen and spores. They also have the great advantage over palynology of not normally requiring extensive laboratory preparation work. On the other hand, palynology comes into its own when analysing borehole sequences, since such fossils can be prepared from small diameter cores and even chippings; cores at least 10 cm wide normally need to be used if plant macrofossils are to be identified reliably.

For correlations between palaeoareas of the same palaeokingdom, things become a little more difficult. As has been shown, most palaeoareas have their own biozonal schemes, partly as a consequence of the endemic taxa in each phytochorion. Even where there are common taxa between the palaeoareas, there may be discrepancies in their ranges. As pointed out by Laveine *et al.* (1989), this is mainly controlled by the speed of geographical dispersal of particular plant groups; pteridophytic plants that produced isospores capable of wind dispersal (i.e. most ferns, sphenophytes and lycophytes) could spread out much more quickly than seed plants. They were able to demonstrate significant discrepancies, for instance, in the first occurrences of certain trigonocarpalean pteridosperms between the Europe and Cathaysia palaeoareas in the Carboniferous, which they put down mainly to the large size of the seeds. Consequently, attempts at correlation between palaeoareas should be based, at least initially on groups such as the ferns and sphenophytes.

Long-range, global correlations between palaeokingdoms are virtually impossible using plant macrofossils alone. This is simply because each palaeokingdom, by definition, shares very few taxa with the others. Where form-genera are reported in common (e.g. the marattialean fern *Asterotheca* has been reported from both the Eurameria and Gondwana palaeokingdoms) great care has to be taken that they do indeed represent the same type of plant, and not just a morphological homologue. The detection of major changes in vegetation triggered by global climatic shifts, such as at the mid-Carboniferous and Palaeophytic–Mesophytic boundaries, might allow the broad correlation of a few stratigraphic levels. However, it is difficult to be certain that such vegetational changes are even approximately synchronous at different latitudes.

An analysis of the biostratigraphy in ecotones might offer a means of correlating biozonal schemes of different palaeokingdoms. Some examples of 'mixed assemblages', such as the presence of *Glossopteris* leaves in otherwise typical Cathaysian (Archangelsky & Wagner 1983) and Angaran (Zimina 1967) assemblages, may reflect long-distance migrations of individual species rather than ecotones in the strict sense. However, the Kazakhstania Palaeoarea in the Lower Carboniferous appears to be a true ecotone, between the Eurameria and Angara palaeokingdoms. Even when such an opportunity is available, though, it may not necessarily allow detailed inter-palaeokingdom correlations; the ecotone, being at the very limit of the geographical range of the palaeokingdoms, may well demonstrate atypical biostratigraphic ranges.

Finally, it is worth emphasizing the importance of environmental factors in biostratigraphic analysis. This has not always been given sufficient weight, and can cause serious bias in the results. One of the best documented cases is in the Upper Carboniferous of the Eurameria Palaeokingdom, with differences in biostratigraphic ranges observed in the adpression and coal-ball assemblages. In some cases, such as the base of the range of callistophyte foliage (*Dicksonites* when preserved as adpressions), it is at about the same stratigraphic level in both facies (middle Westphalian D). In others, however, such as the neuropterid trigonocarpalean foliage, there can be major discrepancies. According to the model outlined by DiMichele *et al.* (1985), this is due to the trigonocarpaleans having mainly occupied areas with clastic substrates, and thus being particularly prone to being preserved as adpressions. Only during periods when the climate became somewhat drier did they migrate into the coal-forming swamps, and thus become preserved within coal balls.

REFERENCES

Anderson, J. M. (1977) The biostratigraphy of the Permian and Triassic. Pt. 3 — A review of Gondwana Permian palynology with particular reference to the northern Karroo Basin, South Africa. *Mem. Bot. Surv. South Africa* **41** 1–67.

Anderson, J. M. & Anderson, H. M. (1985) *Palaeoflora of southern Africa. Prodromus of South African megafloras Devonian to Lower Cretaceous*. Balkema, Amsterdam.

Archangelsky, S. (1984) Floras Neopaleozoicas del Gondwana y su zonación estragigráfica. Aspectos paleogeográ ficos conexos. *Comun. Serv. Geol. Portugal* **70** 135–150.

Archangelsky, S. & Azcuy, C. L. (1985) Carboniferous palaeobotany and palynology in Argentina. *C. R. 10e Congr. Int. Strat. Géol. Carb.* (Madrid, 1983) **4** 267–280.

Archangelsky, S. & Cúneo, R. (1984) Zonacion del Permico continental de Argentina sobre la base de sus plantas fosiles. *Actas III Congr. Latinoamer. Paleont.* (Oaxtepec, 1984) 143–153.

Archangelsky, S. & Wagner, R. H. (1983) *Glossopteris anatolica* sp. nov. from uppermost Permian strata in south-east Turkey. *Bull. Br. Mus. nat. Hist.* (Geol.) **37** 81–91.

Archangelsky, S., Azcuy, C. L., Pinto, I. D., González, C. R., Marques Toigo, M., Rössler, O. & Wagner, R. H. (1980) The Carboniferous and Early Permian of the South American Gondwana area: a summary of biostratigraphic information. *Actas II Congr. Argentino Paleont. Bioestrat.* (Buenos Aires, 1978) **4** 257–269.

Archangelsky, S., Azcuy, C. L. & Wagner, R. H. (1981) Three dwarf lycophytes from the Carboniferous of Argentina. *Scripta Geol.* **64** 1–35.

Assama, K. (1959) Systematic study of so-called *Gigantopteris. Sci. Rep., Tohoku Univ.* 2nd Ser. (Geology) **31**, 1–72.

Barthel, M. (1976) Die Rotliegendflora Sachsens. *Abh. Staatl. Mus. Mineral. Geol.* **24** 1–190.

Bell, W. A. (1938) Fossil flora of Sydney Coalfield. *Mem. Geol. Surv. Canada* **215** 1–334.

Bertrand, P. (1914) Les zones végétales du terrain Houiller du Nord de la France. Leur extension verticale par rapport aux horizons marins. *Ann. Soc. Géol. N.* **43** 208–254.

Bertrand, P. (1919) Les zones végétales du terrain Houiller du Nord de la France. *C. R. Acad. Sci, Paris* **168** 780–782.

Boulter, M. G., Spicer, R. A. & Thomas, B. A. (1988) Patterns of plant extinction from some palaeobotanical evidence. In: Larwood, G. P. (ed.) *Extinction and survival in the fossil record.* Systematics Association, Oxford (Special Volume No. 34), pp. 1–36.

Broutin, J., Doubinger, J., Farjanel, G., Freytet, P., Kerp, H., Langiaux, J., Lebreton, M.-L., Sebban, S. & Satta, S. (1990) Le renouvellement des flores au passage Carbonifère Permien: approches stratigraphique, biologique, sédimentologique. *C. R. Acad. Sci. Paris* **311** 1563–1569.

Chandra, S. & Surange, K. R. (1979) Revision of the Indian species of *Glossopteris. B. Sahni Inst. Palaeobot. Monogr.* **2** 1–291.

Cleal, C. J. (1978) Floral biostratigraphy of the upper Silesian Pennant Measures of South Wales. *Geol. J.* **13** 165–194.

Cleal, C.J. (1984) The Westphalian D floral biostratigraphy of Saarland (Fed. Rep. Germany) and a comparison with that of South Wales. *Geol. J.* **19** 327–351.

Cleal, C. J. (1986) Fossil plants of the Severn Coalfield and their biostratigraphic significance. *Geol. Mag.* **123** 553–568.

Cleal, C. J. (1987) Macrofloral biostratigraphy of the Newent Coalfield, Gloucestershire. *Geol. J.* **22** 207–217.

Clement-Westerhof, J. A. (1984) Aspects of Permian palaeobotany and palynology. IV. The conifer *Ortiseia* Florin from the Val Gardena Formation of the Dolomites and the Vicentian Alps (Italy) with a revised concept of the Walchiaceae (Göppert) Schimper. *Rev. Palaeobot. Palynol.* **41** 51–166.

Corsin, P. & Corsin, P. (1971) Zonation biostratigraphique du Houiller des bassins du Nord-Pas-de-Calais et de Lorraine. *C. R. Acad. Sci., Paris (Sér. D)* **273** 783–788.

Daemon, R. F. & Quadros, L. P. (1970) Bioestratigrafia do Neopaleozoico da Bacia do Paraná. *An. XXIV Congr. Bras. Geol., Soc. Bras. Geol.* (Brasilia, 1970) 359–412.

DiMichele, W. A., Phillips, T. L. & Peppers, R. A. (1985) The influence of climate and depositional environment on the distribution and evolution of Pennsylvanian coal-swamp plants. In: Tiffney, B. H. (ed.) *Geological factors and the evolution of plants*. Yale University Press, Harvard, pp. 223–256.

DiMichele, W. A., Phillips, T. L. & Olmstead, R. G. (1987) Opportunistic evolution: abiotic environmental stress and the fossil record of plants. *Rev. Palaeobot. Palynol.* **50** 151–178.

Dix, E. (1934) The sequence of floras in the Upper Carboniferous, with special reference to South Wales. *Trans. R. Soc. Edinb.* **57** 789–838.

Doubinger, J. (1956) Contribution à l'étude des flores autuno-stéphaniennes. *Mém. Soc. Géol. France* N.S. **75** 1–180.

Durante, M. V. (1976) Paleobotanicheskoe obocnovanie stratigrafii Karbona i Permi Mongolii. *Trudy Sovmestn. Sov.-Mongol. Exped.* **19** 5–279 [in Russian].

Eagar, R. M. C. (1956) Additions to the non-marine fauna of the Lower Coal Measures of the north-Midlands coalfields. *L'pool Manchr Geol. J.* **1** 328–369.

Fissunenko, O. P. & Laveine, J.-P. (1984) Comparaison entre la distribution des principales espèces-guides végétales du Carbonifère moyen dans le bassin du Donetz (URSS) et les bassins du Nord-Pas-de-Calais et de Lorraine (France). *C. R. 9e Congr. Int. Strat. Géol. Carbon.* (Washington & Urbana, 1979) **1** 95–100.

Florin, R. (1938–1945) Die Koniferen des Oberkarbons und des unteren Perms. *Palaeontographica* B **85** 1–730.

Gao L. (1984) Carboniferous spore assemblages in China. *C. R. 9e Congr. Int. Strat. Géol. Carbon.* (Washington & Urbana, 1979) **2** 103–108.

Gomankov, A. V. & Meyen, S. V. (1986) Tatarinovaya flora (sostav i rasprostranenie v pozdnei Permi Evrazii). *Trudy Geol. Inst. Akad. Nauk. SSSR* **401** 1–173 [In Russian].

Gorelova, S. G., Men'shikova, L. V. & Khalfin, L. L. (1973) Fitostratigrafiya i opredelitel' pactenii verkhnepaleozoiskikh uglenosnykh otlozhenii Kuznetskogo Besseina. *Trudy S.N.I.I.G.G.I.M.Ca.* **140** 1–169 [In Russian].

Gothan, W. (1952) Der 'Florensprung' und die 'Erzgebirgische Phase' Kossmatts. *Geologie* **11** 41–49.

Graizer, S. G. (1967) *Nizhnekamennougol'nye otlozheniya Sayano-Altaiskoi skladchatoi oblasti.* Nauka, Moscow [in Russian].

Gu & Zhi (1974) *Palaeozoic plants from China.* Scientific Press, Beijing [In Chinese].

Halle, T. G. (1927) Palaeozoic plants from Central Shansi. *Palaeontol. Sinica Ser. A* **2** 1–316.

Haubold, H. (1980) Die biostratigraphische Gliederung des Rotliegenden (Permosiles) im mittleren Thüringer Wald. *Schriftenr. Geol. Wiss.* **16** 331–356.

Hedberg, H. D. (ed.) (1976) *International stratigraphic guide*. Wiley, New York.

Holland, C. H., Audley-Charles, M. G., Bassett, M. G., Cowie, J. W., Currey, D., Fitch, F. J., Hancock, J. M., House, M. R., Ingham, J. K., Kent, P. E., Morton, N., Ramsbottom, W. H. C., Rawson, P. F., Smith, D. B., Stubblefield, C. J., Torrens, H. S., Wallace, P. & Woodland, A. W. (1978) *A guide to stratigraphic procedure*. Geological Society of London (Special Report No. 11).

Kemp, E. M., Balme, B. E., Helby, R. A., Kyle, R. A., Playford, G. & Price, P. L. (1977) Carboniferous and Permian palynostratigraphy in Australia and Antarctica: a review. *Austral. Geol. Geophys. Jour.* **2** 177–208.

Kerp, J. H. F. (1988) Aspects of Permian palaeobotany and palynology. X. The west- and central European species of the genus *Autunia* Krasser emend Kerp (Peltaspermaceae) and the form-genus *Rhachiphyllum* Kerp (callipterid foliage). *Rev. Palaeobot. Palynol.* **54** 249–360.

Kerp, J. H. F. & Fichter, J. (1985) Die Makrofloren dessaarpfä lzischen Rotliegenden (? Ober-Karbon– Unter Perm; SW-Deutschland). *Mainzer Geowiss. Mitt.* **14** 159–286.

Kozur, H. (1978) Beiträge zur Stratigraphie des Perms. Teil III(1): Zur Korrelation der überwiegend kontinentalen Ablagerungen des obersten Karbons und Perms von Mittel- und Westeuropa. *Freiberger Forschungsh. C* **342** 117–142.

Laveine, J.-P. (1986) La flore du bassin Houiller du Nord de la France. Biostratigraphie et méthodologie. *Ann. Soc. Géol. N.* **106** 87–93.

Laveine, J.-P., Zhang S. & Lemoigne, Y. (1989) Global paleobotany, as exemplified by some Upper Carboniferous pteridosperms. *Bull. Soc. Belge Géol.* **98** 115–125.

Li X. (1963) Fossil plants of the Yuenmenkou Series, North China. *Acta Palaeont. Sinica* N.S. **6** 1–185 [In Chinese].

Li X. (1964) The succession of Upper Palaeozoic plant assemblages of North China. *C. R. 5e Congr. Int. Strat. Géol. Carbon.* (Paris, 1962) **2** 531–537.

Li X. & Yao Z. (1985) Carboniferous and Permian floral provinces in East Asia. *C. R. 9e Congr. Int. Strat. Géol. Carbon.* (Washington & Urbana, 1979) **5** 95–101.

Lyons, P. C. & Darrah, W. C. (1989) Earliest conifers of North America: upland and/or paleoclimatic indicators. *Palaios* **4** 480–486.

Maiboroda, A. A. (1979) O stratigraficheskom raschlenenii Kamennougol'nykh kontinental'nykh otlozhenii tsentral'nogo Kazakhstana. *C. R. 8e Congr. Int. Strat. Géol. Carbon.* (Moscow, 1975) **3** 288–292 [In Russian].

Mamay, S. H. (1986) New species of Gigantopteridaceae from the Lower Permian of Texas. *Phytologia* **61** 311–315.

Mamay, S.H. (1989) *Evolsonia*, a new genus of Gigantopteridaceae from the Lower Permian Vale Formation, north-central Texas. *Amer. J. Bot.* **76** 1299–1311.

Mamay, S. H., Miller, J. M., Rohr, D. M. & Stein, W. E. (1988) Foliar morphology and anatomy of the gigantopterid plant *Delnortea abbottiae*, from the Lower Permian of west Texas. *Amer. J. Bot.* **75** 1409–1433.

Mei M. (1984) *The analysis of the floras of Permian coal-bearing strata in Fujian, Jiangxi and Sichuan provinces*. Graduate School, China Institute of Mining, Beijing.

Meyen, S. V. (1982) The Carboniferous and Permian floras of Angaraland. (A synthesis). *Biol. Mem.* **7** 1–109.

Meyen, S. V. (ed.) (1983) *Paleontologicheskii atlas Permskikh otlozhenii Pechorskogo ugol'nogo basseina*. Komi Branch of the Geological Institute, Leningrad (in Russian).

Meyen, S. V. (1987) *Fundamentals of palaeobotany*. Chapman & Hall, London.

Neuburg, M. F. (1948) *Verkhnepaleozoiskaya flora Kuznetskogo Basseina. Geological Institute, Moscow (Paleontologiya SSSR,* **12-3-2**) [In Russian].

Neuburg, M. F. (1961) Present state of the question on the origin, stratigraphic significance and age of Paleozoic floras of Angaraland. *C. R. 4e Congr. Int. Strat. Géol. Carbon.* (Heerlen, 1958) **2** 443–452.

Novik, E. O. (1952) *Kamennougol'naya flora evropeiskoi chasti SSSR*. Palaeontological Institute, Moscow (*Paleontologiya SSSR,* New Series, Vol. 1) [In Russian].

Novik, E. O. (1978) *Flora i fytostratigrafiya verkhnego Karbona severnogo Kavkaza*. Institute of Geological Sciences, Kiev (in Russian).

Ouyang S. (1982) Upper Permian and Lower Triassic palynomorphs from eastern Yunnan, China. *Can. J. Earth Sci.* **19** 68–80.

Owens, B. (1984) Miospore zonation of the Carboniferous. *C. R. Congr. Int. Strat. Géol. Carbon*. (Washington & Urbana, 1979) **2** 90–102.

Paproth, E., Dusar, M., Bless, M. J. M., Bouckaert, J., Delmer, A., Fairon-Demaret, M., Houlleberghs, E., Laloux, M., Pierart, P., Somers, Y., Streel, M., Thorez, J. & Tricot, J. (1983) Bio- and lithostratigraphic subdivisions of the Silesian in Belgium, a review. *Ann. Soc. Géol. Belg.* **106** 241–283.

Phillips, T. L. (1980) Stratigraphic and geographic occurrences of permineralized coal-swamp plants — Upper Carboniferous of North America and Europe. In: Dilcher, D. L. & Taylor, T. N. (eds) *Biostratigraphy of fossil plants*. Dowden, Hutchinson & Ross, Stroudsburg, pp. 25–92.

Phillips, T. L. (1981) Stratigraphic occurrences and vegetational patterns of Pennsylvanian pteridosperms in Euramerican coal swamps. *Rev. Palaeobot. Palynol.* **32** 5–26.

Phillips, T. L., Peppers, R. A. & DiMichele, W. A. (1985) Stratigraphic and interregional changes in Pennsylvanian coal-swamp vegetation: environmental inferences. *Int. J. Coal Geol.* **5** 43–109.

Radchenko, M. I. (1961) Paleophytological basis for the stratigraphy of the Carboniferous of Kazakhstan. *C. R. 4e Congr. Int. Strat. Géol. Carbon*. (Heerlen, 1958) **2** 559–562.

Radchenko, M. I. (1967) *Kamennougol'naya flora yugo-vostochnogo Kazakhstana*. Satpaev Institute of Geological Sciences, Alma-Ata [In Russian].

Radchenko, M. I. (1979) Kamennougol'naya flora Kazakhstana. *C. R. 9e Congr. Int. Strat. Géol. Carbon*. (Moscow, 1975) **3** 298–301 [In Russian].

Radchenko, M. I. (1985) *Atlas (opredelitel') Kamennougol'noi flory Kazakhstana*. Satpaev Institute of Geological Sciences, Alma-Ata [In Russian].

Raymond, A. (1985) Floral diversity, phytogeography, and climatic amelioration during the Early Carboniferous (Dinantian). *Paleobiology* **11** 293–309.

Read, C. B. & Mamay, S. H. (1964) Upper Paleozoic floral zones and floral provinces of the United States. *U.S. Geol. Surv., Prof. Pap.* **454** K1–K35.

Retallack, G. (1978) Floral ecostratigraphy in practice. *Lethaia* **11** 81–83.

Retallack, G. (1980) Late Carboniferous to Middle Triassic megafossil floras from the Sydney Basin. *Bull. geol. Surv. NSW.* **94** 384–430.

Rigby, J. F. (1985) Aspects of Carboniferous palaeobotany in eastern Australia. *C. R. 10e Congr. Int. Strat. Gol. Carbon.* (Madrid, 1983) **4** 307–312.

Roberts, J. (1985) Australia. In: Wagner, H., Winkler Prins, C. F. & Granados, L. F. (eds) *The Carboniferous of the world. II. Australia, Indian Subcontinent, South Africa, South America, & North Africa.* Instituto Geológico y Minero de España, Madrid, 9–145.

Rocha Campos, A. C. & Archangelsky, S. (1985) South America. In: Wagner, H., Winkler Prins, C. F. & Granados, L. F. (eds) *The Carboniferous of the world. II. Australia, Indian Subcontinent, South Africa, South America, & North Africa.* Instituto Geológico y Minero de España, Madrid, pp. 175–297.

Rössler, O. (1978) The Brasilian eogondwanic floral succession. *Bol. Inst. Geocêinc. Univ. Sao Paulo* **9** 91-95.

Salmenova, K. Z. (1979) Osobennosti Permskoi flori yzhnogo Kazakhstana i ee svyazi s sosednimi florami. *Paleontol. Zhurn.* **4** 119–127 [In Russian].

Schopf, J. M. & Askin, R. A. (1980) Permian and Triassic floral biostratigraphic zones of southern land masses. In: Dilcher, D. L. & Taylor, T. N. (eds) *Biostratigraphy of fossil plants*. Dowden, Hutchinson & Ross, Stroudsburg, pp. 119–152.

Schweitzer, H.-J. (1986) The land flora of the English and German Zechstein sequences. In: Harwood, G. M. & Smith, D. B. (eds) The English Zechstein and related topics. *Geol. Soc. Lond., Spec. Publ.* **22** 31–54.

Scott, A. C. & Chaloner, W. G. (1983) The earliest fossil conifer from the Westphalian B of Yorkshire. *Proc. R. Soc. Lond. B* **220** 163–182.

Scott, A. C., Galtier, J. & Clayton, G. (1984) Distribution of anatomically preserved floras in the Lower Carboniferous in Western Europe. *Trans. R. Soc. Edinb., Earth Sci.* **75** 311–340.

Stoneley, H. M. M. (1958) The Upper Permian flora of England. *Bull. Br. Mus. Nat. Hist. (Geol.)* **3** 295–337.

Surange, K. R. (1975) Indian Lower Gondwana floras: a review. In: Campbell, K. S. W. (ed.) *Gondwana geology*. Australia National Univ. Press, Melbourne, pp. 135–147.

Sutherland, P. K. & Manger, W. L. (eds) (1984) *Neuvième Congrès International de Stratigraphie et de Géologie du Carbonifère. Washington and Urbana-Champaigne May 17-26, 1979. Compte rendu Volume 2 Biostratigraphy*. Southern Illinois University Press, Carbondale and Edwardsville.

Tripathi, C. & Singh, G. (1985) Carboniferous flora of India and its contemporaneity in the world. *C. R. 10e Congr. Int. Strat. Géol. Carbon.* (Madrid, 1983) **4** 295–306.

Vakhrameev, V. A., Dobruskina, I. A., Meyen, S. V. & Zaklinskaja, E. D. (1978)

Paläozoische und mesozoische Floren Eurasiens und die Phytogeographie dieser Zeit. Gustav Fischer, Jena (Russian edition published in 1970 by the Geological Institute USSR, Moscow).

Visscher, H., Kerp, J. H. F. & Clement-Westerhof, J. A. (1986) Aspects of Permian palaeobotany and palynology. VI. Towards a flexible system of naming Palaeozoic conifers. *Acta Bot. Neerl.* **35** 87–99.

Wagner, R. H. (1984) Megafloral zones of the Carboniferous. *C. R. 9e Congr. Int. Strat. Géol. Carbon*. (Washington & Urbana, 1979) **2** 109–134.

Wagner, R. H., Fernandez Garcia, L. G. & Eagar, R. M. C. (1983) *Geology and palaeontology of the Guardo Coalfield (NE León–NW Palencia), Cantabrian Mts*. Instituto Geológico y Minero de España, Madrid.

Wang Z. (1989) Permian gigantic palaeobotanical events in North China. *Acta Palaeontol. Sinica* **28** 314–343 [In Chinese and English].

Yang S., Li X. & Gao L. (eds) (1983) China. In: Wagner, R. H., Winkler Prins, C. F. & Granados, L. F. (eds) *The Carboniferous of the world. I. China, Korea, Japan & S.E. Asia*. Instituto Geológico y Minero de España, Madrid, pp. 10–172.

Zeiller, R. (1894) Sur les subdivisions du Westphalien du Nord de la France d'après les caractères de la flore. *Bull. Soc. Géol. Fr.* **22** 483.

Zhang L. (ed.) (1987) *Carboniferous stratigraphy in China*. Science Press, Beijing.

Zhao X. & Wu X. (1985) Carboniferous macrofloras of south China. *C. R. 9e Congr. Int. Strat. Géol. Carbon*. (Washington & Urbana, 1979) **5** 109–114.

Zimina, V. G. (1967) On *Glossopteris* and *Gangamopteris* found in Permian deposits of Southern Primoriye. *Paleont. J.* **1** 98–106.

Zodrow, E. L. & Cleal, C. J. (1985) Phyto- and chronostratigraphic correlations between the late Pennsylvanian Morien Group (Sydney, Nova Scotia) and the Silesian Pennant Measures (south Wales). *Can. J. Earth Sci* **22** 1465–1473.

6

Palaeoclimatology

C. J. Cleal

Climate influences most aspects of the formation of sedimentary rocks, from the initial production of the sediment by erosion, through its transport and distribution, to its final deposition and diagenesis. Clearly, understanding the climatic conditions prevailing at the time is important if the geological history of the Palaeozoic is to be properly understood. Abiotic signals, such as palaeosols, and the distribution of coals, evaporites and aeolian sediments have been widely used to try to understand climatic trends at this time (e.g. Bless *et al.* 1984, Besly 1986). There have also been attempts to extrapolate climatic conditions from atmospheric and oceanic circulation patterns, inferred from the continental plate configurations (e.g. Rowley *et al.* 1985).

With the known sensitivity of Recent plants to climatic influence, it might be thought that palaeobotanic evidence would be a powerful tool in understanding Palaeozoic palaeoclimates. However, there are major problems in interpreting the evidence; Palaeozoic vegetation consisted of significantly more primitive groups of plants than even the Mesozoic, and it is far from certain that uniformitarian principles can be applied to their responses to climate. The potential importance of the subject makes it impossible not to discuss it briefly in this book. However, because of the inherent problems of interpretation, the following chapter is much shorter and less conclusive than its predecessors.

6.1 CLIMATIC SIGNALS FROM PLANT MORPHOLOGY

6.1.1 Leaf physiognomy

This refers to specific morphological characteristics of leaves which might reflect a climatic influence. The subject has been discussed in a palaeontological context by Spicer (1990) and Chaloner & Creber (1990). However, much of the debate has been based around angiosperm fossils from the upper Mesozoic and Cenozoic, where it is relatively easy to correlate specific characteristics of the fossils with climate by examining Recent plants. In the Palaeozoic, where the fossils mostly represent plants only very distantly related to anything growing today, the situation is less straightforward. Nevertheless, there may be some features which provide a climatic signal.

6.1.1.1 Leaf size
In Recent plants, there is a very broad correlation between leaf size, and temperature and humidity — tropical, humid conditions favour plants with larger leaves. In the Carboniferous and Permian, this correlation still seems to hold good, with the remains of the largest leaves (e.g. of the trigonocarpalean pteridosperms; Laveine 1986) occurring in the tropical Eurameria Palaeokingdom. In the higher palaeolatitudes, smaller leaves tend to predominate, such as of the arberialeans (*Glossopteris*) in the southern hemisphere and the 'cordaites' (*Cordaites*, *Rufloria*) in the north (the higher palaeolatitude cordaite leaves are significantly smaller than the tropical ones; e.g. Meyen 1966, Rothwell 1988). However, there are problems with using leaf size as a climatic index. For instance, there can be a strong taphonomic bias, due to sorting during transport (Spicer 1981). Also, trying to estimate the size of some of the more complex compound leaves or fronds, based on fragmentary fossils, can be difficult (Cleal & Shute 1991).

6.1.1.2 Leaves with drip-tips
Many Recent plants growing in tropical rain forests have leaves which develop an elongate tip, known as a drip-tip (Sporne 1974). It is thought to facilitate the run-off of rain from the leaf-surface. Leaves with what may be drip-tips have been described from the Carboniferous Eurameria Palaeokingdom, such as in the lagenostomalean form-genus *Mariopteris* (e.g. Kidston 1925). However, this was probably the leaf of a liana, and the possibility cannot be ruled out that the elongate tips of the foliage were actually to assist the plant in clinging to the tree around which it was growing.

6.1.1.3 Xeromorphic leaves
Xeromorphy (characters to restrict water-loss from the leaves) are most clearly seen in the cuticles, e.g. thick cuticles, sunken stomata, numerous epidermal hairs. The term implies that it is an adaptation to drier habitats. However, the problem of water-conservation can be just as severe for plants growing in some wet environments, such as where the substrate is saturated by low-pH or mineral-rich solutions. It may also be caused by plants having their leaves in more exposed, windy positions, in which excess transpiration may occur despite the presence of humid atmosphere. Great care has therefore to be taken in interpreting xeromorphy in the fossil record in terms of climate.

6.1.2 Tree rings
The growth rings found in some woods are an indication of variation in rates of growth, usually associated with seasonal changes in weather (Creber & Chaloner 1984, 1985). Consequently, tree rings are associated with plants growing in temperate climates, while plants growing in tropical, more or less seasonless, climates generally have no rings. This is essentially the pattern recognizable in the Palaeozoic, with woods from the Eurameria Palaeokingdom usually being without rings, while woods from the Angara Palaeokingdom generally have distinct rings (Meyen 1987).

In principle, it is possible to determine details of climatic variability from tree rings. Thick rings tend to develop in years when growing conditions were favourable, and thin rings reflect years of poor growth. Clearly, therefore, uniform ring thicknesses tend to suggest that conditions were uniform from year to year. There has been considerable interest in this as a means of determining climatic cyclicity in the Quaternary, but there has been little attempt to use it in the Palaeozoic.

6.2 CLIMATIC SIGNALS FROM VEGETATIONAL CHANGES

In the Cenozoic, it is possible to draw some conclusions about climate from the taxonomic composition of plant fossil assemblages. This is most accurately achieved using the nearest living relative (NLR) method outlined by Axelrod & Bailey (1969). This uses the temperature tolerances of the nearest living relative of each taxon to estimate the temperature tolerance of the vegetation represented by the fossil assemblage. This is difficult in the Mesozoic, because of the phylogenetic distance between the fossil plants and anything living today, while in the Palaeozoic it becomes impossible to draw any meaningful conclusions. Nevertheless, changes in the composition of Palaeozoic plant fossil assemblages may provide some useful evidence of climate change, which may help to corroborate palaeozoological or abiotic evidence.

6.2.1 Climatic amelioration in the Early Carboniferous

Raymond (1985) and Raymond *et al.* (1985) have reported an apparent reduction in palaeophytogeographical provincialism in the uppermost Lower Carboniferous, based largely on polar ordination analyses of the fossil record. This is put down to an amelioration of the climate, resulting from shifts in oceanic currents caused by the fusion of the Laurasia and Gondwana palaeocontinents (whether this fusion of the two plates was as early as the early Namurian is probably doubtful).

Some of the palaeophytogeographical distinctions recognized in these studies are, to say the least, subtle. There are also problems with the data set used, such as the erroneous inclusion of some Upper Carboniferous assemblages from Gondwana. It is clearly important, therefore, that the data be corrected and preferably expanded before too much weight is given to these results. If they can be confirmed, however, they will have to be integrated with possible floristic changes that may be connected with the onset of the palaeoantarctic glaciation (see section 6.2.3, below).

6.2.2 Late Carboniferous climatic fluctuations in the tropical belt

Detailed compositional analyses of the Upper Carboniferous coal-ball assemblages of the Eurameria Palaeokingdom have revealed consistent patterns of changes (see section 5.1.1.2). It has been argued that they reflect major vegetational changes within the tropical swamp forests caused by alternating wet and dry climatic periods (e.g. Phillips & Peppers 1984, DiMichele *et al.* 1985). Their conclusions are based on (a) differences in the assumed tolerances of the various types of arborescent lycophytes in the forest to wet conditions; and (b) the presence or absence of tree-ferns and cordaites, which are assumed to have favoured drier conditions. What they are actually detecting is mainly variations in wetness of the substrate in which the plants were growing, which might be caused by factors other than climate (e.g.

topographical changes altering water-table height). However, from the evidence that they have uncovered so far, a climatic origin for the variation is probably the most convincing.

6.2.3 Climatic changes associated with Late Carboniferous glaciation

Significant changes in plant fossil assemblages, known as the floral break in the Eurameria Palaeokingdom (the base of the *Pecopteris aspera* Biozone) and the base of the 'Ostrogian cooling episode' in the Angara Palaeokingdom (base of the *Rhodeopteridium yavorskii* Biozone), are probably related to the onset of the major glaciation in the mid-Carboniferous (Wagner 1982, Meyen 1987). The establishment of the glossopterid-dominated assemblages in the topmost Carboniferous or basal Permian of the Gondwana Palaeokingdom almost certainly reflects a significant retreat of the palaeoantarctic ice-sheets. It is important to bear in mind, however, that, although palaeobotany appears to be providing evidence of major floristic changes that are coeval with major climatic changes, the plant fossils on their own do not prove what the climatic changes were; virtually all that we know about the history of the late Carboniferous glaciation is derived from abiotic evidence (John 1979). Far more will need to be known about the plants which produced the Palaeozoic fossils before they can be used directly as palaeoclimatic indicators.

6.3 INFLUENCE OF PALAEOZOIC VEGETATION ON PALAEOCLIMATE

This last part of the chapter will follow a different and more speculative philosophy from the rest of the book. It will discuss the possibility that changes in vegetation, particularly in the tropics, may have influenced the climatic changes that took place towards the end of the Palaeozoic. Even if a definite correlation could be found between these events, it would not make the plant fossils geological 'tools', as they are in a biostratigraphic, palaeogeographic or palaeoecologic sense. However, by obtaining a better understanding of the very late Palaeozoic vegetation, we might obtain a better insight into what might have triggered the climatic changes that had such a significant impact on the biotic and abiotic evolution of Earth. There is here also a possible model for present-day and future climatic changes that might result from major destruction of tropical vegetation.

It has been widely recognized that tropical vegetation underwent a significant change and probable decline during the Late Carboniferous. Except in the Cathaysia palaeocontinent, virtually all of the lycophyte/fern-dominated forests had disappeared by the Early Permian. The traditional view has been that the destruction of these forests was a consequence of the climatic aridification of the tropical areas. According to models such as that outlined by Rowley *et al.* (1985), this climatic shift was due to the formation of the Pangea 'super-continent', which disrupted atmospheric and oceanic flows. In particular, it is thought to have instigated a monsoonal pattern of atmospheric circulation that caused the equatorial palaeolatitudes to dry out as the main rainfall was displaced to the north and south.

There is an alternative scenario, however, originating from arguments advanced by Remy (1980) that the tropical forests were destroyed by topographic changes lowering the water-table, rather than a decrease in precipitation. The forests started to decline in the very late Westphalian, and this probably coincided with the collision between Laurasia and Gondwana. The collision would have produced a degree of topographic uplift in the areas where the forests grew, producing the lowering of the water-table required by the Remy model. Initially, the forests were able to survive as the dominant tree-forms changed from lycophytes to ferns, which had a greater tolerance to drier conditions (DiMichele *et al.* 1985). By the start of the Permian, however, even the ferns were unable to survive the progressive drying of the substrate. Only in the Cathaysia palaeocontinent, where the effects of this collision were not felt, could lycophyte-dominated forests continue to flourish during the Permian.

What replaced the fern-dominated forests of tropical Pangea in the Early Permian is far from clear. Stands of conifers were certainly in existence, as fragments of foliage occur periodically in the fossil record, but did they form tracts of forest of similar extent to the Late Carboniferous tropical forests? Meyen (1987) argued, based on the abundant pollen found in strata of this age, that there were indeed extensive conifer forests at this time. Even accepting that much of the tropical Permian sediments was unsuitable for the preservation of plant fossils, however, it is remarkable how meagre the plant fossil record is in the Lower Permian. Also, if extensive forests were in existence in the tropics at this time, some evidence of the soils would be expected to be found; such evidence is extremely sparse. It is likely, therefore, that the tropical forests of Pangea were to a large extent destroyed in the Early Permian.

There is, therefore, the following possible scenario. Topographic changes in the Late Carboniferous resulted in the large-scale destruction of the tropical forests. As a consequence, the take-up of CO_2 from the atmosphere was reduced, and this resulted in a raising of temperatures, as well as a shift in rainfall patterns. This led to the aridification of tropical Pangea, and may also have instigated the melting of the palaeoantarctic ice-caps, which seems to have happened at about the same time. Temperate regions also seem to show evidence of climatic warming during the very late Carboniferous and Early Permian, both in the southern and northern hemispheres. During most of the Permian, climatic conditions seem to have been stable, but towards the end of the period (*c.* 30 million years after the destruction of the tropical forests) there was a catastrophic worldwide disruption of vegetation, widely referred to as the Palaeophytic–Mesophytic transition.

This model may be too simplistic. The truth probably lies somewhere between it and the types of models proposed by Rowley *et al.* (1985) and Besly (1986), where climatic change is induced by changes in continental configuration. The important point is that vegetational changes in the tropics had at least some impact on climate towards the end of the Palaeozoic. There are of course a number of important questions that need answering. We still do not know enough about the population dynamics of the forests and how they would have reacted to ecological change. What was the proportional destruction of the tropical forests? Could the destruction of the tropical forests have any connection with the much later catastrophic changes at the very end of the Permian? There is clearly much scope for future work.

These changes are obviously important for understanding the evolution of the Earth and its biota during the late Palaeozoic. However, they may also have a more immediate significance. There is much talk today of climatic changes being caused by the destruction of the modern tropical forests, through the so-called 'greenhouse effect'. Much of the discussion is based on computer simulations of the global climate, which, to say the least, are difficult things to produce. In the Upper Palaeozoic, however, we may have a model which shows what *actually* happened when large areas of tropical forest were destroyed. The model is not an exact fit for Recent conditions; the forests consisted of more primitive plants, the continental configurations were different, and the time-scales appear to have been longer. Nevertheless, it something that actually happened, rather than just an 'imagined' scenario in the memory of a computer, and is worthy of much closer scrutiny when trying to forecast the future of our planet.

REFERENCES

Axelrod, D. I. & Bailey, H. P. (1969) Paleotemperature analysis of Tertiary floras. *Palaeogeog. Palaeoclimat. Palaeoecol.* **6** 163–195.

Besly, B. M. (1986) Sedimentological evidence for Carboniferous and Early Permian palaeoclimates of Europe. *Ann. Soc. Géol. N.* **106** 131–143.

Bless, M. J., Bouckaert, J. & Paproth, E. (1984) Migration of climatic belts as a response to continental drift during the Late Devonian and Carboniferous. *Bull. Soc. Belge Géol.* **93** 189–195.

Chaloner, W. G. & Creber, G. (1990) Do fossil plants give a climatic signal? *J. Geol. Soc. Lond.* **147** 343–350.

Cleal, C. J. & Shute, C. H. (1991) The Carboniferous pteridosperm frond *Neuropteris heterophylla* (Brongniart) Sternberg. *Bull. Br. Mus. Nat. Hist. (Geol.)* **46** 153–174.

Creber, G. T. & Chaloner, W. G. (1984) Influence of environmental factors on the wood structure of living and fossil trees. *Bot. Rev.* **50** 357–448.

Creber, G. T. & Chaloner, W. G. (1985) Tree growth in the Mesozoic and early Tertiary and the reconstruction of palaeoclimates. *Palaeogeog. Palaeoclimat. Palaeoecol.* **52** 35–60.

DiMichele, W. A., Phillips, T. L. & Peppers, R. A. (1985) The influence of climate and depositional environment on the distribution and evolution of Pennsylvanian coal-swamp plants. In: Tiffney, B. H. (ed.) *Geological factors and the evolution of plants.* Yale University Press, New Haven, pp. 223–256.

John, B. S. (1979) The Great Ice Age: Permo-Carboniferous. In: John, B. S. (ed.) *The winters of the world. Earth under the ice ages.* David & Charles, Newton Abbott, pp. 154–172.

Kidston, R. (1925) Fossil plants of the Carboniferous rocks of Great Britain. Part 6. *Mem. Geol. Surv. G.B., Palaeont.* **2** 523–670.

Laveine, J.-P. (1986) The size of the frond in the genus *Alethopteris. Geobios* **19** 49–56.

Meyen, S. V. (1966) *Cordaiteans of the Upper Paleozoic of North Eurasia.* Geological Institute, Acad. Sci. USSR, Moscow [In Russian].

Meyen, S. V. (1987) *Fundamentals of palaeobotany.* Chapman & Hall, London.

Phillips, T. L. & Peppers, R. A. (1984) Changing patterns of Pennsylvanian coal-swamp vegetation and implications of climatic control on coal occurrence. *Int. J. Coal Geol.* **3** 205–255.

Raymond, A. (1985) Floral diversity, phytogeography, and climatic amelioration during the Early Carboniferous (Dinantian). *Paleobiology* **11** 293–309.

Raymond, A., Parker, W. C. & Parrish, J. T. (1985) Phytogeography and paleoclimate of the Early Carboniferous. In: Tiffney, B. H. (ed.) *Geological factors and the evolution of plants.* Yale University Press, New Haven, pp. 169–222.

Remy, W. (1980) Wechselwirkung von Vegetation und Boden im Paläophytikum. In: *Festschrift far Gerhard Keller.* Wenner, Osanbrück, pp. 43–79.

Rothwell, G. W. (1988) Cordaitales. In: Beck, C. B. (ed.) *Origin and evolution of gymnosperms.* Columbia University Press, New York, pp. 273–297.

Rowley, D. B., Raymond, A., Parrish, J. T., Lottes, A. L., Scotese, C. R. & Ziegler, A. M. (1985) Carboniferous paleogeographic, phytogeographic and paleoclimatic reconstructions. *Int. J. Coal Geol.* **5** 7–42.

Spicer, R. A. (1981) The sorting and deposition of allochthonous plant material in a modern environment at Silwood Lake, Silwood Park, Berkshire, England. *U.S. Geol. Surv., Prof. Pap.* **1143** 1–77.

Spicer, R. A. (1990) Climate from plants. In: Briggs, D. E. G. & Crowther, P. R. (eds) *Palaeobiology. A synthesis.* Blackwell, Oxford, pp. 401–403.

Sporne, K. R. (1974) *The morphology of angiosperms.* Hutchinson, London.

Wagner, R. H. (1982) Floral changes near the Mississippian–Pennsylvanian boundary: an appraisal. In: Ramsbottom, W. H. C., Saunders, W. B. & Owen, B. (eds) *Biostratigraphic data for a mid-Carboniferous boundary.* IUGS Subcommission on Carboniferous Stratigraphy, Leeds, pp. 120–127.

7

Concluding remarks

C. J. Cleal

Plant fossils should be treated no differently from other types of palaeontological material; they should be studied with the goal of extracting the maximum of information relevant to the biological and geological sciences. It is hoped that this book has shown that plant fossils can provide information of geological as well as biological significance. They cannot always compete with marine animal fossils (e.g. graptolites, ammonoids) for biostratigraphic resolution, but within non-marine facies they can often be the best tools available. Their palaeoecological value can also be considerable, in the right facies, provided that the problem of transport can be resolved. For palaeogeographic studies, however, they are unrivalled, at least from the Lower Carboniferous upwards.

It is, of course, important not to divorce the geological and biological aspects of the study of plant fossils. The geological information that they supply is based mainly on patterns of their distribution. This is controlled, at least partly, by when and where the plants originally grew and thus, in turn, on their biology. Equally, much of the biological information that can be derived from plant fossils can only be accurately interpreted when seen in its geological context (stratigraphic, palaeogeographic, palaeoecological). It gives the lie to the view that palaeobotany is merely a branch of botany, to be pursued by botanists. It is a speciality in its own right, that requires skills taken from both the botanical and geological sciences. In an ideal and financially wealthy world, there would be separate schools or departments of palaeobotany, where the particular blend of skills could be appropriately nurtured. One such school actually exists (the Birbal Sahni Institute at Lucknow), but for practical reasons the vast majority of palaeobotany is done in departments of either botany or geology. It is hoped that this book will be of particular help to the potential palaeobotanists approaching the subject from the latter direction.

It was stated in Chapter 1 that one purpose of the book was to provide a model for pursuing the geological investigation of plant fossils in the Mesozoic and Cenozoic. It would be wrong to say that this aspect of plant fossils has been completely ignored in the higher strata; palaeoecological work has been developed to some degree, and there has also been some analysis of their palaeophytogeography. Their biostratigraphic potential, however, has hardly been developed, other than in some notable studies in the Triassic. The problem is that there has been no coherent attempt to collate the distribution data on plant fossil taxa in the same way as has been done in the Palaeozoic. Without this sort of analysis, any attempt at biostratigraphy or palaeophytogeography will be founded on shifting sand. In principle, though, there is no reason why plant fossils should not be as useful as geological tools in the Mesozoic and Cenozoic as they have proved to be in the Palaeozoic.

Taxonomic index

This is an index of species and form-genera of plant fossils only; families and higher taxa are not included, nor are plant microfossils or animal fossils. For each form-genus, a statement is made as to what type of fossil it represents. This is only intended as a generalized guide, however, and includes heterogenous, descriptive' groups such as pteridosperms.